Lightning strokes. Time exposure 20 sec, aperture f/8. At upper left note intercloud strokes. Illumination behind cloud (sheet lightning) is from stroke at left or from other intercloud strokes. Note intense upward streamer at lower left. Also note branches (forked lightning) that feed charge upward into main stroke. The single bright stroke at the center has two main branch streamers that in turn have branches. The stroke on the right is partly hidden behind a low cloud and appears to have only two branch streamers near the earth which are just barely visible on the original photo. At extreme upper right are more intercloud discharges. It is difficult to discern the particular shapes of the storm clouds, but the central one appears to have more the shape of a bowl that the inverted-bell shape usually ascribed to them. Photograph taken by Bernard Brunette at Lac Simon, Papineauville, Quebec, 1965.

LIGHTNING PROTECTION

Lightning Protection

J. L. MARSHALL

Managing Engineer (Transmission)
Engineering Division
Canadian Broadcasting Corporation

A Wiley-Interscience Publication

JOHN WILEY & SONS, New York · London · Sydney · Toronto

Library of Congress Cataloging in Publication Data:

Marshall, J Lawrence, 1913–
Lightning protection.

"A Wiley-Interscience publication."
1. Lightning protection. I. Title.

TH9057.M37 643 73–4415
ISBN 0–471–57305–1

Printed in the United States of America

10–9 8 7 6 5 4 3 2 1

Preface

Apart from being an interesting natural phenomenon that serves to maintain a balance in the global electrical system, lightning has serious destructive effects on property and life. Any effort to reduce the losses by even a fraction is well worthwhile. However, a search for information on the nature of lightning, its effects, and protection against them would lead to a few specialized books; and many technical papers, each of necessity restricted to particular aspects. One would not likely find a consolidation of the information in one volume that would provide a comprehensive view of the nature of lightning, the principles of protection, and direction toward sources of more analytical and specialized knowledge. With this book I have attempted to fill this gap.

Chapter 1 gives a summary of losses from lightning, to indicate the importance of using protection methods.

Chapter 2 takes a global view of the world's electrical environment and points out how lightning plays a part in a balanced system. It then explains the generation of cloud charge and describes its release in the form of lightning.

The magnitude of the voltage, current, and energy in lightning is given in Chapter 3, and the risk of damage from it is discussed.

The earth is an essential terminal for dissipating lightning energy in protective systems, and Chapter 4 describes this function along with the various ways of coupling to the earth by vertical and horizontal buried conductors.

Chapter 5 outlines the principles of grounding systems and methods of measuring their effectiveness. This is intended to prepare the reader for the design of specific systems.

Chapter 6 discusses specific measures for the protection of human life and describes situations where there is likely to be a risk of danger.

Chapter 7 covers protective systems for communications towers, buildings, and equipment including broadcasting installations. Such installations, owing to their nature and location, are particularly vulnerable to lightning.

Chapter 8 applies the principles of protection to buildings, particularly structures that are isolated, at high elevation, or are themselves very tall. This is becoming increasingly important as the heights of new buildings and towers increase. Systems for hazardous buildings are included as well as methods of home protection.

Chapter 9 outlines the application of grounding principles to electric power systems, including the function and characteristics of high-voltage lightning arresters.

It is my hope that practicing engineers, technical students, and other interested persons will find the book interesting and useful.

J. L. MARSHALL

Montreal, Canada
September 1972

Acknowledgments

I am grateful for the help given by Dr. J. S. Marshall, Director of the McGill Observatory, who read the manuscript and contributed to the section on the physics of cloud formation. Thanks are also given to Mrs. S. C. Sirkett and Mrs. R. Hyland, who patiently typed the manuscript.

J. L. M.

Lightning strokes. Exposure time about 20 sec, aperture f/8. At the upper center is a long forked intercloud discharge. At the upper left is illumination from discharges that are hidden from view. At lower left is a low-intensity stroke to earth. The two bright strokes left of center evidently occurred at different times because they appear to cross each other. They are probably at different distances also. Note the branch streamers that collect charge for the main return stroke. The unusual branch streamer coming from the upper right is possibly a discharge from the upper air or cloud to the more negative lower cloud. Photograph taken by Bernard Brunette at Lac Simon, Papineauville, Quebec, 1965.

Contents

CHAPTER 3

CHAPTER 4

CHAPTER 5

CHAPTER 6

CHAPTER 7

LIGHTNING PROTECTION

CHAPTER 1

The Toll of Lightning

1.1 MAJOR CATEGORIES

The destructive effects of lightning can be segregated into four general classes. Major losses, in North America in particular, are from forest fires caused by lightning. A second category includes the loss of buildings due to fires caused by lightning and physical damage to buildings by rapid thermal expansion induced by a lightning stroke. In a third class we can put the disruption of such public services as electric power, telephone, and communications. Fourth, there is a significant loss of human life due to lightning and, less tragic but nonetheless serious, a substantial loss of livestock.

1.2 LOSSES IN THE UNITED STATES

The destruction of property by lightning in the United States incurs a yearly loss[1] estimated at $100 million. This includes forest fires, building fires, damage to structures, disruption of power lines, aircraft damage, and livestock deaths and injuries.

Lightning is a major cause of farm-building fires and is responsible for about 80% of all livestock losses due to accidents.

A large number of power-supply interruptions are caused each year by lightning. The frequency of transmission-line flashovers varies from one part of the country to another, but the occurrence of 100 flashovers per 100 miles of high-voltage line per year is not uncommon.

The toll in human lives in the United States due to lightning averages about 600 annually, and injuries are inflicted on 1500 others. About four out of five of all deaths and injuries occur in rural areas where people are more often exposed in the open.

1.3 LOSSES IN CANADA

Only partial statistics have been obtained for Canada, and these are mainly related to forest fires. During the 10-year period 1960 to 1969 the average number of lightning-caused fires per year[2] was 2867, and the associated annual loss was nearly $2.5 million. About 80% of these fires were in forests.

Canadian data[3] on human deaths caused by lightning show that ten persons died in 1969.

1.4 LOSSES IN GREAT BRITAIN

Two aspects of loss statistics for Britain have been referred to by Golde.[4] With regard to human casualties, there are about 10 people killed annually by lightning in England and Wales.

The second statistical item is that the number of direct lightning strokes that cause interruptions is about 20 per 100 miles per year on power-transmission lines of 132-kV rating, the number of strikes to lines of lower voltage, which are lower in height and have shorter crossarms, is smaller.

1.5 LOSSES IN GENERAL

The frequency of thunderstorms varies from one geographic location to another, the highest frequency being in the equatorial zone. If the statistics of the United States, for example, are extrapolated to encompass the temperate and tropical zones of the earth and their corresponding populations, the toll of lightning in the world would be some 6000 lives annually and property losses of about $1 billion.

Occasionally there has been damage caused to historical structures and monuments that cannot be evaluated in monetary terms. More attention is being given to the protection of such structures.

On the positive side lightning, which ionizes the air, produces nitrogen oxide and ozone. Nitrogen oxide is a fertilizer that is beneficial to vegetation. It is even claimed that lightning might have aided the evolution of life on earth. Professor A. T. Oparin of Russia, author of a book called *Origin of Life*,[5] claims that in the early ages of the earth, inorganic carbon-like substances evolved into life aided by exposure to ultraviolet rays, some of which could have come from lightning.

REFERENCES

1. *Lightning,* ESSA/PI 660024, U.S. Department of Commerce, Environmental Science Services Administration, Washington, D.C., December 1966.
2. *Fire Losses in Canada 1965,* Department of Public Works, Fire Protection Branch, Ottawa, Canada.
3. *Causes of Death, Canada, 1969,* Dominion Bureau of Statistics, Ottawa, Canada.
4. R. H. Golde, "Protection of Structures Against Lightning," *Proc. IEE,* **115,** No. 10 (October 1968).
5. A. T. Oparin, *Origin of Life,* Progress Publishers, Moscow.

CHAPTER 2

The Nature of Lightning

2.1 THE GLOBAL ELECTRICAL SYSTEM

A widely accepted view of the global electrical system is that the earth and the lower ionosphere are two highly conductive surfaces separated by an imperfect insulating atmosphere, that is to say, a large condenser with some leakage. Over fair-weather areas there is a downward transfer of positive charge, which tends to reduce the positive potential of the ionosphere and to neutralize the negative charge on the earth. Within the global system lightning discharges transfer positive charge upward at a rate sufficient to sustain a balanced dynamic system; that is, the regular current flow between the positively charged ionosphere and the negatively charged earth is controlled and maintained by global thunderstorm activity. It has been observed that the values of steady fair-weather potential gradient and air–earth current closely follow the thunderstorm diurnal variation curve. This is illustrated in Figs. 2.1 and 2.2. It has also been recorded that solar flares produce increases in the steady electrical field and current flow between ionosphere and earth. Accordingly increased lightning discharges can be expected after such solar outbursts. The ionosphere referred to here is the layer within the height range of 50 to 75 km.

Over the earth's surface as many as 2000 thunderstorms are continually in existence. Active thunderstorms discharge, by lightning, at the average rate of about 20 C every 10 sec, which is equivalent to approximately 2 A of steady current. As the average global air–earth current is on the order of 1500 A, this would indicate the existence of 700 to 800 active storms, and even more if minor storms are included.

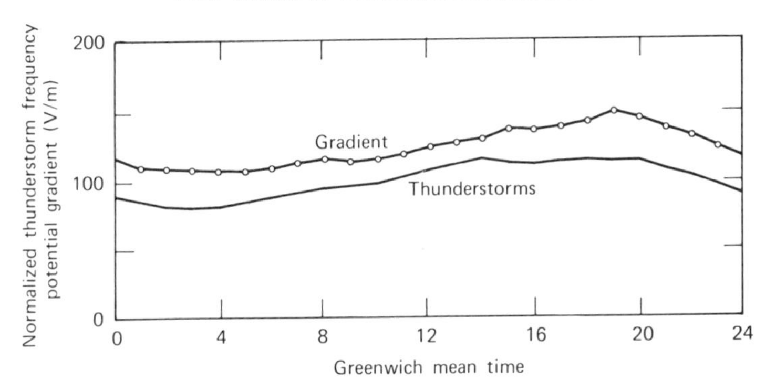

Fig. 2.1. *Diurnal variation of potential gradient and thunderstorm frequency (plotted by Whipple[1]).*

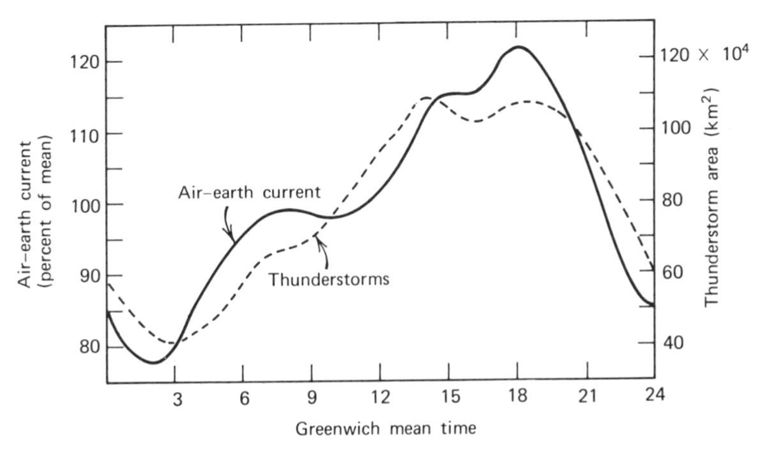

Fig. 2.2. *Diurnal variation of the global thunderstorm area (Whipple[1]) and the fair-weather air–earth conduction current at Mauna Loa Observatory.*

The steady electrical field is about 3 V/cm near ground level under fair-weather conditions. During thunderstorm development this can rise to 500 or 600 V/cm beneath the thunderclouds and to much higher values near ground level below a stepped leader. The total steady potential between ionosphere and earth is estimated at 300 kV. From this and the total steady current of about 1500 A, the resistance of the atmosphere is estimated to be 200 Ω.

2.2 THE GENERATION OF CLOUD CHARGE

The electrical charge that develops in a thundercloud to an intensity sufficient to initiate lightning usually has a concentration of negative charge near the cloud base and a positive charge near its top.

The development of a thundercloud requires air movement, moisture, and a rather specific temperature range. Accordingly thunderstorms often occur on hot summer days when a cold air front approaches.

Electrically the thunderstorm cloud may be regarded as a charge-separation device or electric generator satisfying the "needs" of the global system. Because the air has both ionizing and dielectric properties, the current flow is partly a flow of ions or charged hydrometeors and partly the intense current discharge of lightning. Apart from the vertical flow, there are inter-cloud discharges in a generally horizontal direction—a familiar sight to most observers.

In a mechanical and thermodynamic sense the thunderstorm is a creation of the troposphere in which the temperature decreases with height sufficiently for clouds to build upward buoyantly. These "cumulus" clouds are initiated by condensation of water vapor in excess of equilibrium values, into cloud droplets about 10 μ in diameter. These droplets might freeze into ice crystals, but with or without freezing they aggregate to form precipitation particles whose individual mass is a million times that of the constituent cloud droplets. These precipitation particles are raindrops, snowflakes, graupel (rime-like, low-density mixture of cloud droplets and air), and, least frequently, hail, which has 10 to 100 times more individual mass than the other particles. Raindrops are sometimes melted snowflakes or graupel rather than the first phase or aggregation of cloud droplets. Cumulus clouds can develop one way or another, depending on the degree of the "ice" phase. They can develop into cumulonimbus clouds, which produce rain showers, or into layer-type clouds wherein continuous rain is produced from layers of melting snow. Thunderstorms are of the first category, being rain showers or complexes of rain showers, but not all rain showers become thunderstorms or parts thereof.

In the thermodynamic aspects the phenomenon of supercooling strongly influences the mechanics and electrical properties of rain showers. The thermal behavior of water is also a fundamental factor. Water will not freeze, until it reaches $-40°C$, unless it contains some site on a solid surface from which the crystalline pattern of ice can develop outward. Most water, unless highly purified, does contain solids with such sites on them, which are given the name "nuclei." The greater the number of nuclei, the higher the temperature at which the particle is effective in producing freezing. The rise in temperature is proportional to the exponential increase in the number of nuclei. Every raindrop and cloud droplet is a separate sample requiring its own nucleus if it is to freeze. A raindrop contains enough nuclei to have an effective freezing temperature in the range −15 to −25°C. However, raindrops seldom get this cold. Cloud droplets, possessing fewer nuclei, are seldom active (freezing) at temperatures warmer than −30°C, and many of them would have to reach −40°C before freezing. Unlike larger samples of water, such as those in containers or ponds, which freeze at 1 or 2°C below the nominal freezing point, the cloud droplets require supercooling to freeze. Conversely, to freeze the droplets supercooling must occur, and the freezing process occurs (when it does) over a wide range of temperatures within the cloud and therefore over a wide range of heights.

Cloud droplets and ice crystals fall through still air at speeds of less than 0.3 m/sec; snowflakes fall at 1 m/sec; and raindrops fall at 5 to 10 m/sec. Cumulus clouds are in vertical circulation and contain more or less concentrated updrafts and downdrafts with velocities of 30 m/sec or greater. When the updraft velocity is greater than the fall speed of the raindrops, it is possible for the raindrops to ascend as they grow and to be stored aloft in considerable concentrations, which may reach values of mass such as to neutralize the buoyancy and so turn the updraft into a downdraft.

Cloud growth can be illustrated by the rather familiar example of smoke issuing into the atmosphere. The effluent from the chimney mixes with a larger quantity of the surrounding air, so that what appears to be smoke-filled air has come in large part from the atmosphere, with only a small contribution from the chimney. When the effluent contains much water vapor and little smoke, the white cloud that billows out is, in fact, cloud, made up of cloud droplets condensed from the vapor. A similar process generates cumulus clouds; they start with warm moist air rising from the surface and cooling (because of expansion), then mixing with less moist surrounding air that has been at a height of several kilometers for a period of hours. Clouds are mixtures of air with different histories and include mixtures of a variety of "hydrometeors" (i.e., particles of solid or liquid water); the mixing activity continues through the few-hours' lifetime of the cloud or storm.

Considering now electrical charge generation, there are various ways in which elements of a rain shower may possibly achieve the charge separation that eventually results in lightning. The charge separation is related to the supercooling, and in some cases the freezing, of droplets; and the disposition of charge concentrations in the mature thundercloud is due, in some part, to the vertical circulation (i.e., the updrafts and downdrafts already referred to). An interpretation of particle flow in relation to temperature and height is shown in Fig. 2.3.

The counterflow of air in the developing thundercloud carries a positive charge upward and a negative charge downward. A pictorial representation

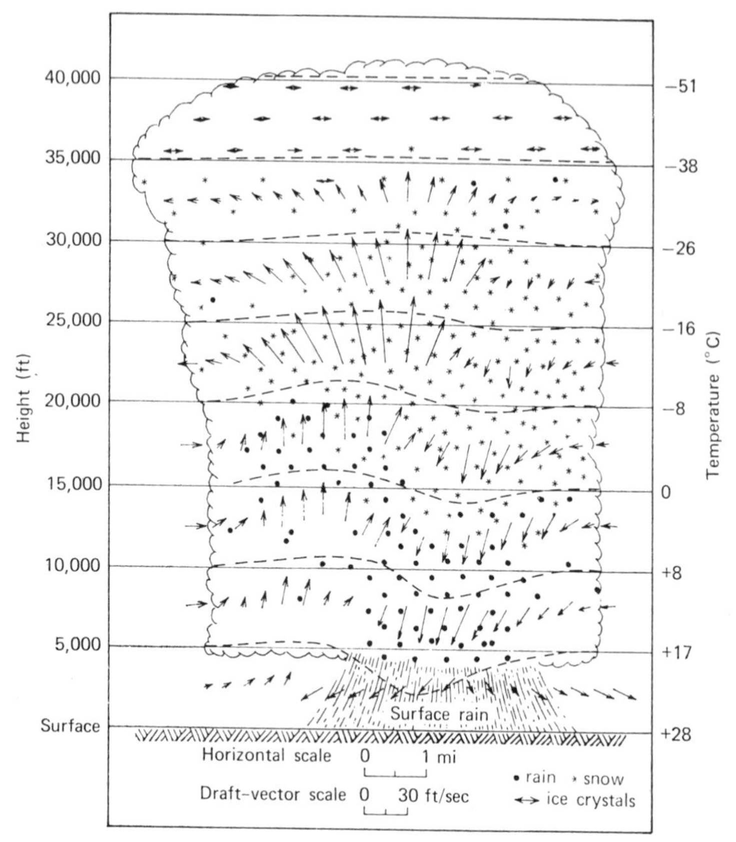

Fig. 2.3. *An idealized cross section through a thunderstorm cell in its mature stage. Key: ●, rain; *, snow; ↔, ice crystals. After Golde.*[2]

of a thundercloud life cycle is shown in Fig. 2.4. This positioning of the charges is sometimes referred to as bipolar, or dipoles of charge.

In addition to the main charge centers, negative in the lower cloud and positive in the upper cloud, pockets of opposite charge accumulate below the lower main charge center and above the upper charge center; that is, a negative layer of charge gathers at the top of the cloud and a positive one at the base. The situation at the mature thundercloud stage is depicted in Fig. 2.5.

A normal thundercloud will comprise several "cells" or dipoles of charge. The whole cloud may have lateral dimensions of several kilometers; its base may be 1 or 2 km above ground, and its top, 10 to 14 km above ground. A photograph of a thundercloud is shown in Fig. 2.6.

Much of the observation of cloud development, electrical field, and current flow has been done by meteorologists, who use weather radar to identify and locate cloud formations. A typical radar map from a ground-based station is shown in Fig. 2.7.

2.3 MATURE STAGE

As the separation of charge proceeds in the cloud cell, the potential difference between the concentrations of charge increases and the potential drop across any vertical unit distance of the charged mass similarly increases. After 20 min or so of the generation process, the cloud will have reached a mature stage, charged to a point where a discharge will be initiated. The temperature at the main negative-charge center will be about $-5°C$ and at the auxiliary pocket of positive charge below it, about 0°C. The main positive-charge center in the upper cloud will be about 15°C colder than its negative counterpart. At the mature stage the total potential difference between the main charge centers will be 10^8 to 10^9 V, and the total stored charge several hundred coulombs. Only a part of the total charge is released by lightning to earth, as there are both intercloud and intracloud discharges as well.

2.4 THE LEADER

There are several varieties of lightning discharge, and of these the dominant cloud-to-earth type will be described first. The channel to earth is first established by a stepped discharge called a leader or leader stroke. The initiation of the leader might be due to the downward movement of negative charge, outside an updraft in the core. Positively charged moisture particles are drawn into this flow, which in turn attracts more negatively charged particles in a funneling action that, under the influence of the strong electrical field, eventually forces a negative streamer out of the base of the cloud into the air. Another possible mechanism is the breakdown between elongated, polarized water droplets at the cloud base caused by the high potential field or a discharge between the negative-charge mass in the lower cloud and the positive pocket of charge below it.

Once in the air the negative streamer advances in steps, seeking areas of positive space charge. It may probe into several branch paths but stop after a short distance in favor of the main channel, which presents more positive charge. The average speed of the stepped leader is about 10^5 m/sec, or

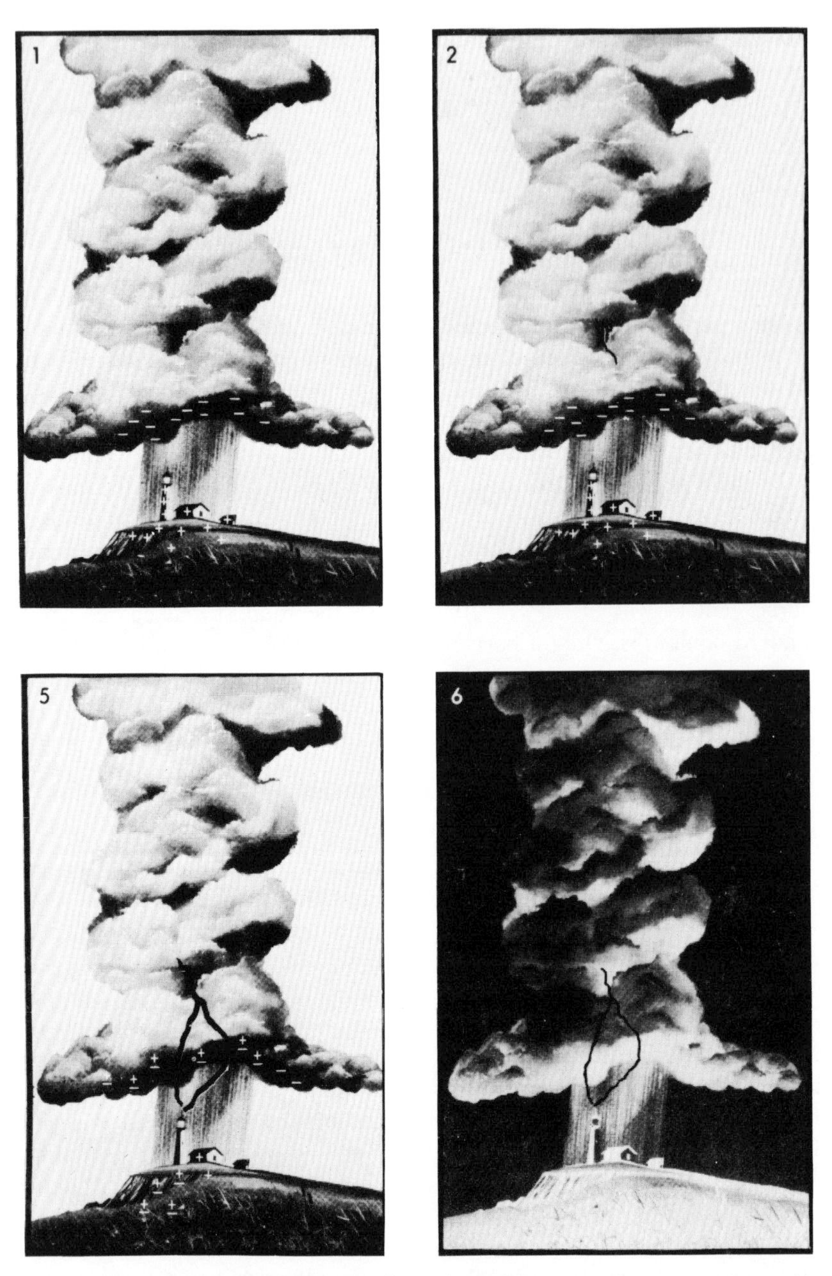

Fig. 2.4. *Life cycle of a lightning stroke.*[3] *As the thunderstorm induces a growing positive charge in earth, the potential between cloud and ground increases (1) until pilot leader starts a conductive channel toward ground (2) and is followed by step leaders (3), which move downward*

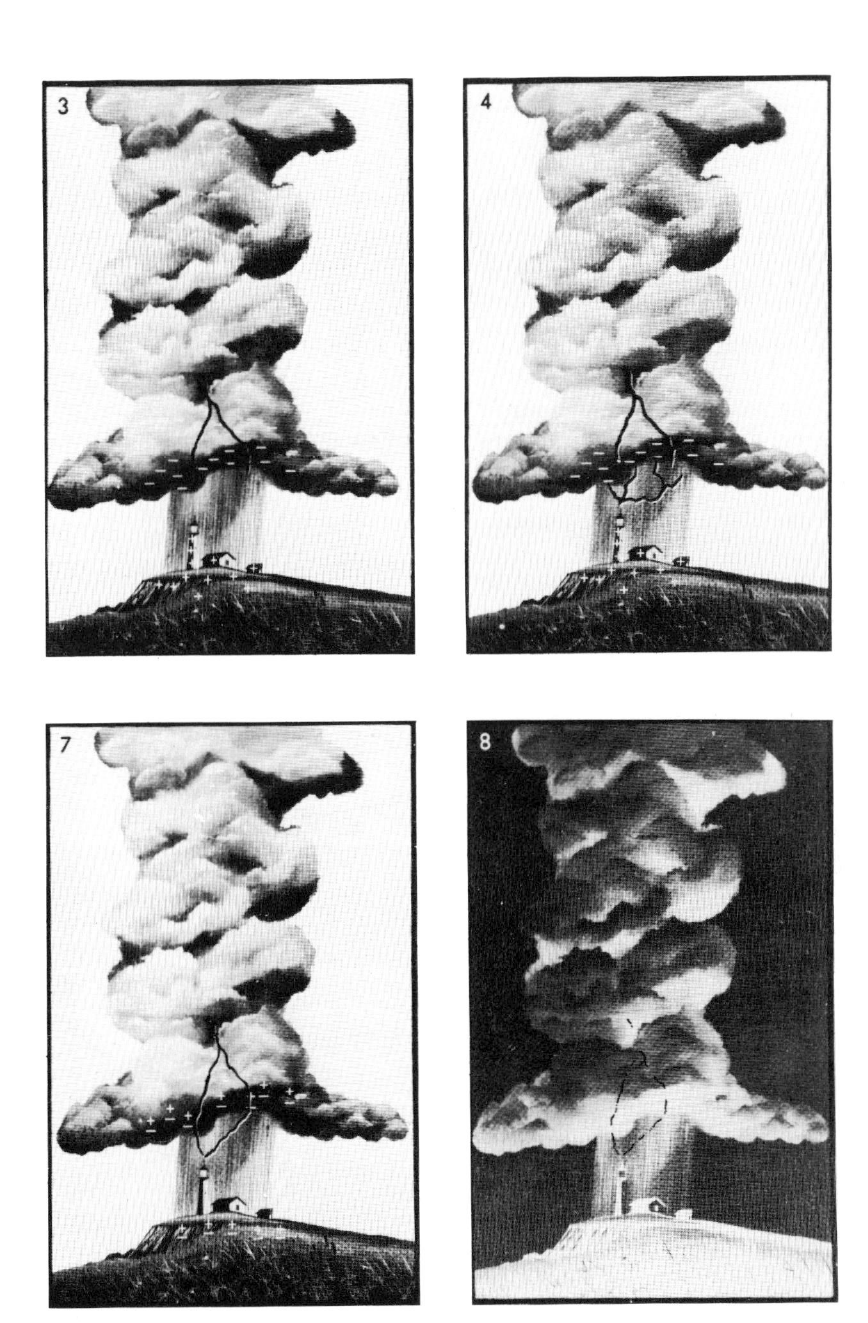

for short intervals (4) until met by streamers from ground. Return stroke from ground illuminates branches (5) and seems to come from cloud. Main stroke is followed by sequence of dart leaders and returns (6 and 7) until the potential is reduced or the ionized path is dispersed (8). Elapsed time: about 1 sec.

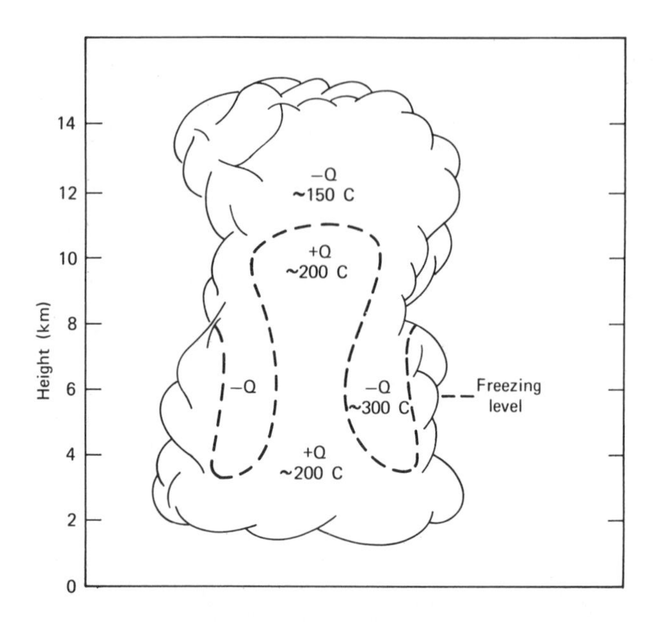

Fig. 2.5. *Estimated charge distribution in a mature thundercloud. After Phillips.*[4]

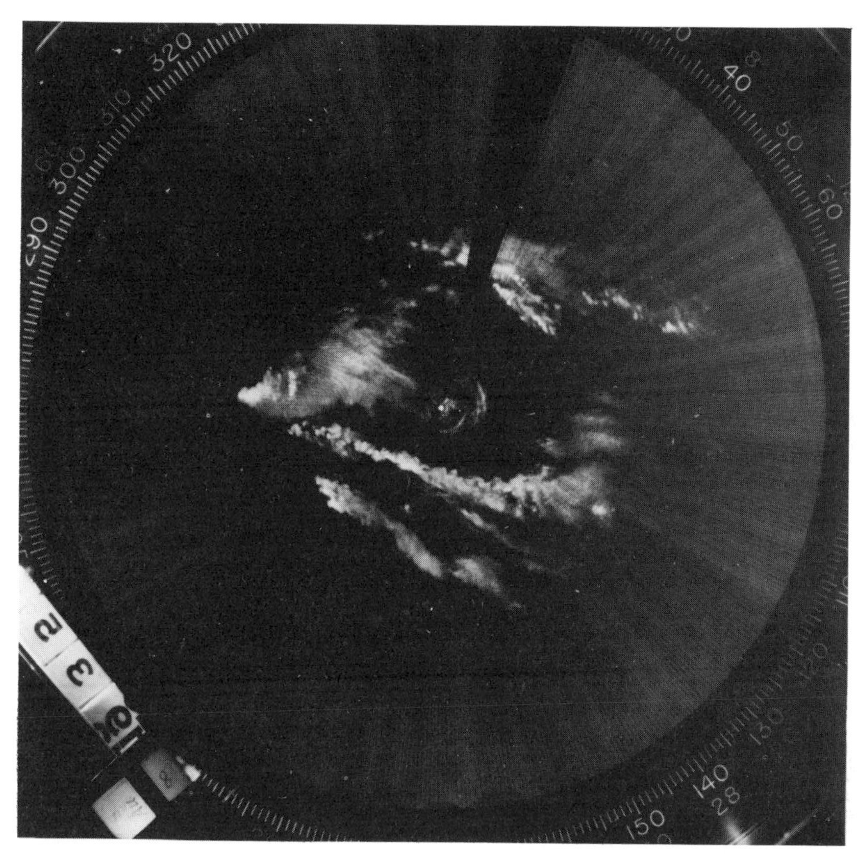

Fig. 2.7. *Radar map of storm clouds. This radar map, produced at the McGill University Radar Weather Observatory near Montreal, shows one line of thunderstorm clouds that has moved through the city and is now about 20 miles south. Another storm, about 40 miles north of the radar, moved south over Montreal 3 hr later. The radar sees only the pattern of precipitation. The smaller intense cells are heavy rain, and the larger hazily defined areas are snow, which melts to light rain before reaching the ground. Electrical charge separation associated with the formation of rain and snow produces lightning discharges.*

Fig. 2.6. *Thundercloud. Rapidly forming cloud turrets from warm moist air that has risen from the ground in the previous 10 min to 1 hr. The air expanded and cooled while rising, and cloud droplets of about 10-μ diameter condensed out, too small to be seen individually, but visible en masse as a cloud. At the top of the cloud the temperature has cooled to about −30°C, but few of the droplets have frozen; and at the top the turrets have mixed in a large amount of the surrounding drier air and have ceased to rise. A fraction of the frozen droplets will grow as soft hail, or graupel, and at this stage a strong electrical field will arise in the surrounding air, indicating that charge separation has occurred. Photograph by Max Sauer of Montreal.*

one-thousandth of the speed of light. Each step of the leader advances its tip a distance of 10 to 200 m, and these spurts are separated by time intervals of 40 to 100 μsec. The tip of the leader bears a corona fringe, and at the completion of each step a pulse of current shoots back toward the cloud. The leader deposits a small portion of the cloud charge along its length which is neutralized by space charge. This amounts to 0.5 to 1 C. The developmental stages of a leader are shown in Fig. 2.8.

While the leader channel is developing there is a displacement current between the charged cloud and ground caused by the high potential difference. This current is supplemented by point discharges from earth objects, such as buildings, towers, trees, or even blades of tall grass. At some instant the concentration of such discharges can constitute an arc that flows upward to meet the leader tip at 20 to 70 m above ground, depending on the existing field potential. The point at which the two channels meet is called the point of strike and is the start of the return stroke.

The stepped leader takes about 20 msec to reach the earth, but the return stroke takes only 100 μsec or thereabouts to flash from earth to cloud.

It is within the last 100 m of ground or of an object thereon that the point of strike is determined. The higher the electrical potential between cloud and ground, the higher will the striking point be. Also the stronger the electrical field, the less likely is the leader to deviate from the vertical as it nears the earth. The area within which a strike may be expected near a tall earthed object is accordingly dependent on the electrical field strength, and this makes it difficult to be specific about the "safety zone" afforded by a tall object.

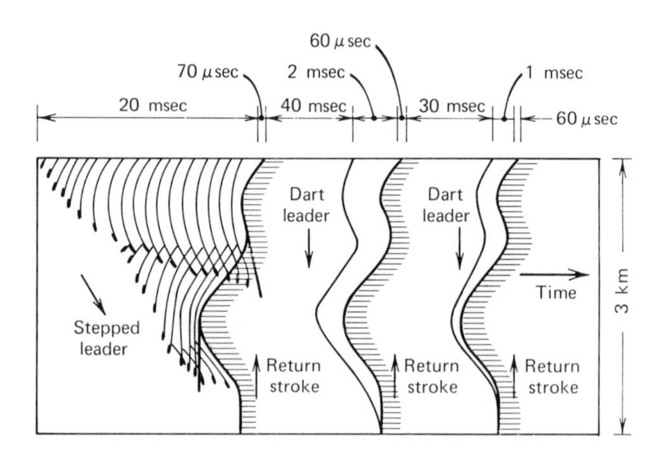

Fig. 2.8. *The developmental stages of a lightning flash. After Uman.*[5]

The "dead-end" branch paths that the stepped leader may have probed feed positive space charge into the return stroke channel, which assumes the potential of the cloud at the instant of strike. These branch paths, or "forks," glow from the heavy current and give rise to the term "forked lightning."

2.5 THE RETURN STROKE

It is the return stroke that has destructive effects and therefore arouses our concern about protection. It can be regarded as an intense positive current *from* ground or as the lowering of negative charge *to* ground. Its purpose is to neutralize the opposite charges between cloud and earth. After the first return stroke it is usual for another region of the cloud to provide sufficient charge for a second stroke or several more, separated by intervals of 10 to 20 msec. This return discharge of one stroke or a succession of strokes is called a flash. The current in a stroke averages about 20 kA, but in exceptionally intense storms it can exceed 100 kA. The average charge released per flash is about 25 C.

The return-stroke current heats the path instantly to temperatures of 15,000 to 20,000°C, making the air luminous and causing the explosive air expansion that we hear as thunder. Each return stroke is a unidirectional current pulse that rises to a crest value in a few microseconds and then decays over a period of several tens or hundreds of microseconds.

The strike point near the earth depends on the strength of the potential gradient below the leader channel and the distribution of space charge in the atmosphere, so that it is not only high objects and elevations alone that receive strikes but also plains, lakes, or even a valley between mountains.

At the instant before a return stroke of the usual kind there is an intense electrical field, with the ground positive and the atmosphere above negative. Immediately after the stroke there is a reversal of field, with the ground becoming negative with respect to a positive space above. This reversal may be explained by the existence of a comparatively large distributed positive charge within the lower portions of the storm cloud.

A strong wind can displace succeeding strokes laterally, making them appear as a wide band known as ribbon lightning.

Some discharges from clouds do not reach the ground. Strokes within clouds often provide general illumination, known as sheet lightning. Sometimes "bead" or "chain" lightning is observed, caused by an intense stepped leader, neutralizing space charge with a high current at each step. Another variety called ball lightning has been observed in the form of a luminous ball moving laterally near the earth, after a nearby lightning discharge. No accepted explanation has yet been given for this variety. Ball lightning is treated in more detail in Section 2.10.

2.6 THE DART LEADER

After the first return stroke, there is usually enough charge in a higher region of the cloud to initiate another leader. Usually this leader follows the path taken by the previous stroke. Owing to the remanent ionization of the path the leader darts to earth directly in about 1 msec. Accordingly it is called a dart leader. The interval between the first return stroke and the dart leader is 70 msec on the average, and thereafter successive dart leaders and their return strokes may recur at 40- to 50-msec intervals. Three or four return strokes along the same path are common. There are as many as 10 return strokes in about 10% of lightning flashes. If a dart leader is too long delayed, due probably to the slowness of charge buildup in the cloud, the path ionization diminishes and the next leader will "step" its way downward, but at a higher velocity and with shorter steps than the original stepped leader.

Whereas the average velocity of the stepped leader is in the range of 0.01 to 0.7% of the velocity of light, the dart leader's velocity is between 0.13 and 10% of the speed of light.

The velocity characteristics of a dart leader compared with those of a stepped leader and with those of a return stroke are illustrated in Fig. 2.8.

2.7 ANALOGY WITH SPARK BREAKDOWN IN AIR

The process of an electrical spark across an air gap has been observed to be similar to a cloud-to-ground lightning discharge,[6] so that the characteristics of the air-gap breakdown are helpful in explaining the lightning discharge. When a sufficiently high voltage is applied across a volume of insulating gas (such as reasonably dry air), the gas breaks down and electrons are released. Above a certain voltage an ionization wave, called a streamer, proceeds from the highly stressed positive electrode toward the cathode, branching out along the way and extending into the gas where the electrical field stress is somewhat less. If the applied voltage is high enough, the vigorous streamer or streamers reach the cathode with a high-potential wavefront. As the wavefront reaches the cathode, electron emission is stimulated by the temporary local electrical field and negative streamers from the cathode are produced. These negative streamers greatly increase the density of ionization in the channel, which yields a "backstroke," or discharge across the gas from cathode to anode.

For longer gas paths or air gaps, the streamers from the anode might be incapable of reaching the cathode. In this case a number of secondary channels will develop at the anode and proceed toward the cathode. The resulting increased ionization of the gas can become dense enough to produce breakdown of the gap and a spark discharge.

2.8 VARIATIONS OF THE LEADER AND STROKE

From mountain peaks or very tall structures, such as the Empire State Building, a positive leader channel may start upward from the peak due to the intense concentration of positive charge accumulated there, forced by the strong electric field. When this leader reaches the cloud, the charge there remains diffuse in the volume of water droplets, with the result that there is no sudden increase in current, but a comparatively gradual current flow of modest amplitude. However, any subsequent strokes return to the common pattern of a downgoing dart leader and a high-current return stroke. The positive leader carries a heavier current than a negative leader (i.e., a few hundred amperes on the average).

Occasionally there is a lightning stroke between a positively charged cloud and ground. The leader for this kind of stroke can be an upward negatively charged leader followed by a positive-current stroke from the cloud or a downward positively charged leader followed by an upward negative-current stroke from the ground.

Multiple lightning strokes or several strokes within a relatively small area can occur, as depicted in Fig. 2.9.

Positive leaders advance without stepping in most cases and at a higher velocity than do negative leaders.

In negative ground-to-cloud strokes the leader also advances upward in steps. The junction point with a positive streamer from the cloud occurs at a considerable height above ground. At high ground altitude the junction has been observed at 1000 to 1800 m above a tower struck by lightning.

2.9 UNUSUAL LIGHTNING DISCHARGES

There are several unusual types of discharge that cause little damage because of their rarity. For example, lightning discharges can be caused by severe dust storms and volcanic eruptions.

At high altitudes, when a thundercloud is passing over, the electrical field can have sufficient intensity to attract a discharge from grounded objects, such as towers or structures, or even a person's hair. This can often be heard

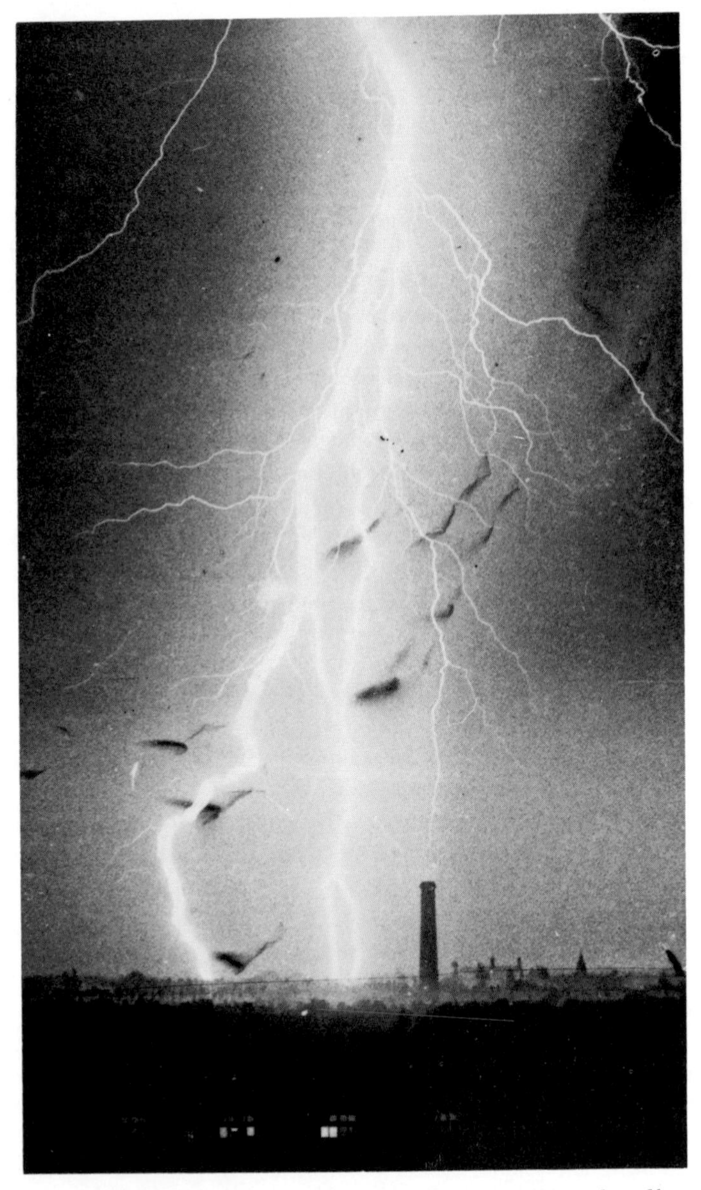

Fig. 2.9. *Lightning flash to earth comprising three strokes. Note the number of branches that also contribute charge to the main return stroke. From Golde.*[2]

as a hissing sound and seen as a glow several inches long if in darkness. The corona current attains values of 1 to 4 μA. It has been estimated that the charge released by such corona discharges annually is from one twenty-fifth to one-seventh the charge released by lightning itself.

Occasionally there is an upward positive streamer from the earth that does not reach the lightning channel; and it might follow in time a distant flash.

Sometimes in urban areas certain kinds of trees, including dead ones, a chimney, or a flagpole will be struck in preference to some adjacent higher tree or other object. An explanation of this phenomenon has been given.[8] It postulates that the susceptible object initially presents a very high resistance to corona-discharge currents flowing between its upper parts and ground. Assuming that the voltage gradient above the object increases under the influence of the charged thundercloud, the leakage current along the object also increases until the "resistor" breaks down due to internal stress, surface flashover, or tracking where there is soot (e.g., along a chimney). The electrical breakdown will suddenly bring the top of the object to earth potential, with a consequent quick rise in voltage gradient between the object and the cloud, which can initiate the lightning discharge. Such occurrences can be avoided by placing, along the height of the object, a down conductor connected to a suitable earth electrode.

2.10 BALL LIGHTNING

The most prevalent and distinctive variant of the lightning discharge is called ball lightning. Its characteristics have been summarized[9] as follows:

1. Usually spherical in shape.
2. Usually occurs toward the end of a nearby electric storm.
3. Varies in size from 10 to 20 cm in diameter, but can be as much as several feet in diameter.
4. Color ranges from brick red to orange to blue.
5. Sometimes accompanied by humming, fluttering, crackling, hissing, or sizzling sounds.
6. Often leaves a smell of ozone.
7. Extinction may be silent or explosive, or simultaneous with a lightning stroke.
8. Exhibits translatory motion by rolling or sliding along metallic objects, such as telephone wires, railway tracks, or wire fences.
9. Has been known to magnetize ferrous objects.

10. Movement is slow, about 2 m/sec, independent of the wind direction and along a tortuous path; it may detach itself from a grounded object and may appear to bounce.
11. Has sufficient energy to melt wire, boil away several pounds of water, burn persons severely, and set fires.
12. Decay can last from several seconds to several minutes.
13. At times dartlike discharges may radiate from the ball.

The variation in point of occurrence and motion of ball lightning is one of the difficulties encountered by analysts who attempt to postulate a theory for it. The discharge can move through dwellings, entering openings like chimneys or windows and sometimes emerging from enclosed spaces; it can move along a section of railway tracks, over or through a vehicle, and so on.

Calculations have indicated that lightning balls can have a surface temperature as high as 5000°C, and they release much energy when they decay.[10] Those that are diffuse and reddish in color decay slowly, whereas a bluish-white ball has a rapid, explosive decay that can cause charring or other destruction.

In addition to ball and surface glow configurations, a large percentage of observations have indicated a doughnut or ring form. This, along with reports of rolling, spinning, or tumbling motion, suggests that it is rotational plasma with possibly an internal ring current.[10] Of the several theories proposed to explain the phenomenon, there are two that receive most credence today.[10] One theory was proposed by Kapitza in 1955, and the second by Watson in 1960.

The Kapitza theory postulates that ionized air and the pressure of vapors form ionized clouds of plasma comprising atomic nuclei stripped of their electrons. They possess their own period of electromagnetic oscillation, which enables them to absorb external electromagnetic energy of the same frequency (i.e., there is a resonant effect). The external dimensions of the ball are critical in determining this resonant absorption of energy. For absorption the external energy wavelength should be 3.65 times the diameter D of the ball,[10] from which the critical volume of the ball should be $\pi D^3/6$. This relationship is supported by observations that lightning balls usually have diameters of 10 to 20 cm, and dominant wavelengths produced in electric storms are 35 to 70 cm.

Watson's theory also includes resonant absorption of external energy by an ionized sphere of plasma occurring when the wavelength of the external radiation is approximately four times the diameter of the ball. The theories differ about the initiation of the ball. Watson associates the formation of the plasma with a node of an electromagnetic field, the aftermath of lightning, whereas Kapitza links the formation of a ball with the antinode.

Another view[10] is that a lightning ball is initially formed immediately after a lightning stroke when the electromagnetic field surrounding the plasma column of the lightning path "pinches" it off to form a ball or "kinks" the ionized column, which bends back on itself to form a ring or doughnut in which current continues to flow. This toroid of plasma is kept rotating by the external electromagnetic field that sustains the ring current.

Experimental work has produced electrodeless discharges simulating ball lightning, and the ability to transfer energy into a remote "ball" possessing light, heat, and electric power might have potential practical applications. With regard to protection, however, the rarity and limited destructive power of ball lightning make special measures unnecessary. This is fortunate because normal lightning protection methods would not be effective.

REFERENCES

1. F. J. W. Whipple, "On Association of the Diurnal Variation of Electric Potential Gradient in Fine Weather with the Distribution of Thunderstorms Over the Globe," *Quart. J. Roy. Meteorol. Soc.*, **55,** No. 229, January 1929.
2. R. H. Golde, "Thunderstorms," *J. IEE,* May 1962.
3. *Lightning,* ESSA/PI 660024, U.S. Department of Commerice, Environmental Science Services Administration, Washington, D.C., December 1966.
4. B. B. Phillips, "Convected Charge in Thunderstorms," *Mon. Weather Rev.*, **95,** No. 12 December 1967.
5. M. A. Uman, *Lightning,* McGraw-Hill, New York, 1969, p. 000.
6. E. Nasser, "Spark Breakdown in Air at a Positive Point," *IEEE Spectrum,* November 1968.
7. B. J. Mason, "Mechanism of the Lightning Flash," *Electronics and Power,* May 1966.
8. J. A. Williams, letter in *Electronics and Power,* November 1969.
9. C. M. Cade and D. Davis, *The Taming of the Thunder Bolts,* Abelard-Schuman, London, 1969.
10. D. J. Ritchie, "Ball Lightning in Nature and in the Laboratory," *JIEE,* May 1963.

BIBLIOGRAPHY

H. R. Byers, *Thunderstorm Electricity,* University of Chicago Press, Chicago.

K. Berger, *J. Franklin Inst. Spec. Issue,* **283,** No. 6, June 1967.

K. Berger, "Front Duration and Current Steepness of Lightning Strokes to the Earth," *Proc. Intern. Conf. Cent. Elect. Res. Labs. May 1962,* Butterworths, London, 1962.

M. A. Uman, *Lightning,* McGraw-Hill, New York, 1969.

CHAPTER 3

Magnitude of the Lightning Discharge

3.1 ACCUMULATED CLOUD CHARGE

There is evidence that a charged thundercloud may possess a total charge of about 1000 C in order to sustain a lightning flash about 20 sec, which is the usual interval. From the measurement of changes in the electrical field at a known distance from the thundercloud producing the lightning discharge a value for the electric moment can be obtained. From these data and an expression for the growth rate of the field while the cloud is regenerating charge, the total cloud charge generated from its beginning can be calculated.

The electrical field growth rate can be expressed[1] as follows:

$$\frac{dE}{dt} + 4\pi\sigma E = 4\pi q v \tag{1}$$

where q = spatial concentration of charge both positive and negative, in coulombs per unit volume,

v = velocity at which these charges are being separated,

σ = conductivity of medium (for leakage currents from cloud in vertical direction).

If E is made zero at the instant when growth of the field starts ($t = 0$), then its rate of growth is

$$E = \frac{qv}{\sigma}(1 - e^{-4\pi\sigma t}) \tag{2}$$

But E can be expressed as $4\pi Q_s/A$, where Q_s is the separated charge during the growth period and A is the cross-sectional area of the cloud mass. Also if Q_g represents the total generated charge since the cloud's inception, the space-charge density q is Q_g/Ah, where h is the height of the space-charge region or the distance between the charge centers in the cloud. This is commonly about 3 km.

Putting $E = 4\pi Q_s/A$ and $q = Q_g/Ah$ into expression 2, we obtain for the total charge

$$Q_g = \frac{2\pi\sigma M}{v(1 - e^{-4\pi\sigma t})} \tag{3}$$

where $M = Q_s h$, the electric moment.

The average value of the time constant, or the interval between flashes, has been observed to be 20 sec, and that of the electric moment M is 110 C-km (coulomb-kilometers).

Then, using 3 km for h, we find that $Q_g \simeq 20{,}000/v$ C. The velocity of charge separation can be taken as 20 m/sec. This gives a value for the total generated charge Q_g of about 1000 C.

This value is in some measure of agreement with the values postulated by Phillips,[2] as depicted in Fig. 2.5. The latter value is predicated on a considerable flow of current, one to several amperes, into the cloud from its environment, plus internal generation of charge. A total diameter of 10 km was assumed for the cloud model, and a total positive-charge accumulation of 400 C, distributed in two equal parts within the central updraft column. The opposing negative charge is made up of about 150 C flowing inward over the cloudcap from above and about 300 C generated by a downward current flow associated with precipitation, which accumulates in the lower part of a cylindrical sheath about the central positive core.

3.2 MAGNITUDE OF SEVERAL TYPES OF STROKE

Although the lightning stroke that lowers a negative charge to earth is the most common, it is interesting to compare its quantitative characteristics with those of the less frequent positive-charge-lowering stroke, as done by Berger,[3] whose data are reproduced in Fig. 3.1.

Interesting data on the upward-going streamer that rises to meet a downward positive stroke are shown in Table 3.1. The downward positive stroke, which is more common to high-altitude terminals, will be of more frequent concern as more very tall (1000 ft or more) towers and buildings are built.

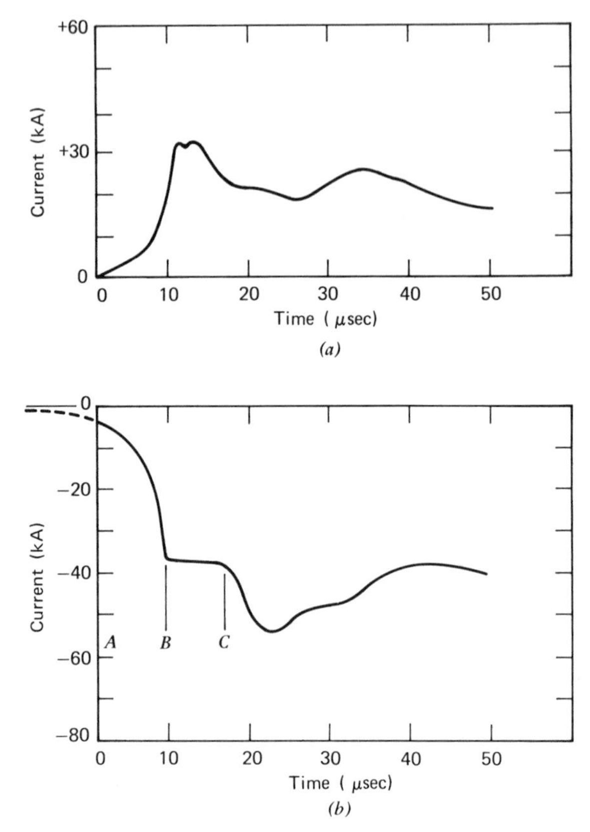

Fig. 3.1. *Comparison of the steepnesses of usual negative strokes and a positive downward stroke without connecting streamer. (a) Downward positive stroke without connecting streamer. (b) Usual downward negative stroke. Up to point A is the formation of the very faint "streamer discharges" from the positive tower. In the interval from A to B there is formation of a highly conducting bridge between downcoming and upward growing "streamer discharge." Point C marks secondary current initiation from branching. After Berger.*[3]

3.3 VALUES FOR THE COMMON STROKE IN DETAIL

The detailed quantitative aspects of the usual negative-cloud-to-positive-earth lightning stroke have been compiled by Uman.[4] They are listed in Table 3.2.

It will be noted that in a representative storm the stepped leader channel will accrue a charge of about 5 C, which will in a later instant be partially released to earth by a (return) lightning stroke. Ensuing dart leaders will add

TABLE 3.1 CHARACTERISTICS OF SOME "UPWARD CONNECTING STREAMERS" TO SIX DOWNWARD POSITIVE STROKES[a, b]

Tower No.	T^{c} (msec)	H^{d} (m)	Q^{e} (C)	i^{f} (kA)	$(di/dt)_{max}{}^{g}$ (kA/μsec)
2	0	0	30	32	17
1	3.0	500	12	22	4.5
2	6.1	1000	62	77	3
2	8.7	1200	65	56	2
2	11.6	1150	35	27	1
2	14	1800	130	106	2

[a] Adapted from Berger.[3]
[b] Tower is negative electrode.
[c] Time from start of upward leader until beginning of impulse current i.
[d] Vertical length of upward leader up to its arrival at the existing lightning channel into the cloud (velocity is considered to be constant above this length).
[e] Charge of impulse current within 2 msec after start of the impulse.
[f] Impulse current (peak value).
[g] Steepness (tangent to the current-time curve) of the impulse current.

about 1 C each to the leader channel, and the following two or three return strokes will lower an additional 20 C or more to earth, so that the lightning flash (sum of the strokes) will transfer about 25 or 30 C of charge to earth. It will also be seen that the maximum values of current and charge per flash are four to five times greater than the representative or median value.

3.4 STEADY CONDUCTION CURRENTS

The potential difference between the charged thundercloud and earth required to initiate a stepped leader is on the order of 10^8 V.

At this stage there are also leakage conduction currents from the cloud, upward to the electrosphere (lower ionosphere), transferring positive charge, and downward, transferring negative charge to earth. The upward flow of positive charge, on approaching the lower ionosphere (about 50 km), spreads round the earth in a relatively short time.[5]

The earth normally bears a negative charge that stays near the value of 1.1×10^{-9} C/m^2. It has been estimated[5] that the average conduction current flowing in fair-weather areas in the form of negative charges upward to the positive electrosphere and positive charges downward to the earth is 3×10^{-12} A/m^2, which, taken over the whole surface of the earth, amounts

TABLE 3.2 QUANTITATIVE ASPECTS OF COMMON CLOUD-TO-EARTH LIGHTNING STROKES[a]

	Minimum	Representative	Maximum
Stepped leader			
Length of step (m)	3	50	200
Time interval between steps (μsec)	30	50	125
Average velocity of propulsion (m/sec)	1.0×10^5	1.5×10^5	2.6×10^6
Charge deposited on channel (C)	3	5	20
Dart leader			
Velocity of propulsion (m/sec)	1.0×10^6	2.0×10^6	2.1×10^7
Charge deposited on channel (C)	0.2	1	6
Return Stroke			
Velocity of propulsion (m/sec)	2×10^7	5×10^7	1.4×10^8
Rate of current increase (kA/μsec)	1	10	80
Time to peak current (μsec)	1	2	30
Peak current (kA)		10–20	110
Time to fall to half peak current (μsec)	10	40	250
Charge transferred, excluding continuing current (C)	0.2	2.5	20
Channel length (km)	2	5	14
Lightning flash			
Number of strokes per flash	1	3–4	26
Interval between strokes in absence of continuing current (msec)	3	40	100
Duration of flash (sec)	10^{-2}	0.2	2
Charge transferred, including continuing current (C)	3	25	90

[a] Indicative values, compiled by Uman[4] from many sources.

to a total of approximately 2000 A. Other estimates have been in the range of 1500 to 1800 A. Since these current values are derived from measurements of air conductivity and potential gradients, it is not surprising to see some variation in the results. Using the same method, the steady current flowing from a developed thundercloud and the electrosphere has been found to range from 0 to 1.4 A, averaging 0.8 A.

The equivalent steady current transferred between clouds and earth per storm[6] is in the range 0.3 to 1 A. Accordingly for an energy balance to be maintained one would expect about 2000 storms to occur per day.

3.5 ENERGY IN LIGHTNING

To estimate the amount of energy in a typical lightning flash let us take a value of 10^7 V for the breakdown voltage between cloud and ground and assume a total discharge of 20 C. Then the energy released is 20×10^7 J, or about 55 kW-hr, in the one or more strokes that make up the flash. The crest current per stroke, which may rise in 1 or 2 μsec and fall to half its value in 40 or 50 μsec, is about 20 kA on the average, but it can reach several times this value in an intense storm.

The energy of the flash dissipated in the air channel is expended in several processes. Small amounts of energy produce dissociation of molecules, ionization, excitation of the earth's magnetic field, kinetic energy of the channel particles (about 1%), and radiation (about 0.4%). A large part of the energy, about 98%, is consumed in the sudden expansion of the air channel. Some fraction of this total causes heating and, at times, fracture of the earth or grounded object that is struck. In general lightning returns to the global system the heat energy that originally created the charged cloud.

The stroke-channel expansion occurs at supersonic speed and produces a shock wave, roughly cylindrical in shape, which we hear as thunder. The temperature of the channel may rise to 30,000°K. After the shock wave the pressure equilibrium is restored in a few tens of microseconds, and the channel then expands more slowly.

3.6 LIGHTNING-STROKE CURRENTS

Figure 3.1 shows the positive- and negative-current waveforms of downward strokes, with descriptive notes from Berger.[3] The current waveforms of the initial discharge of a number of strokes are shown in Fig. 3.2.

The overall duration of a lightning stroke, including leader time, is about 70 msec, and that of a flash comprising three or four strokes, about 0.25 sec.

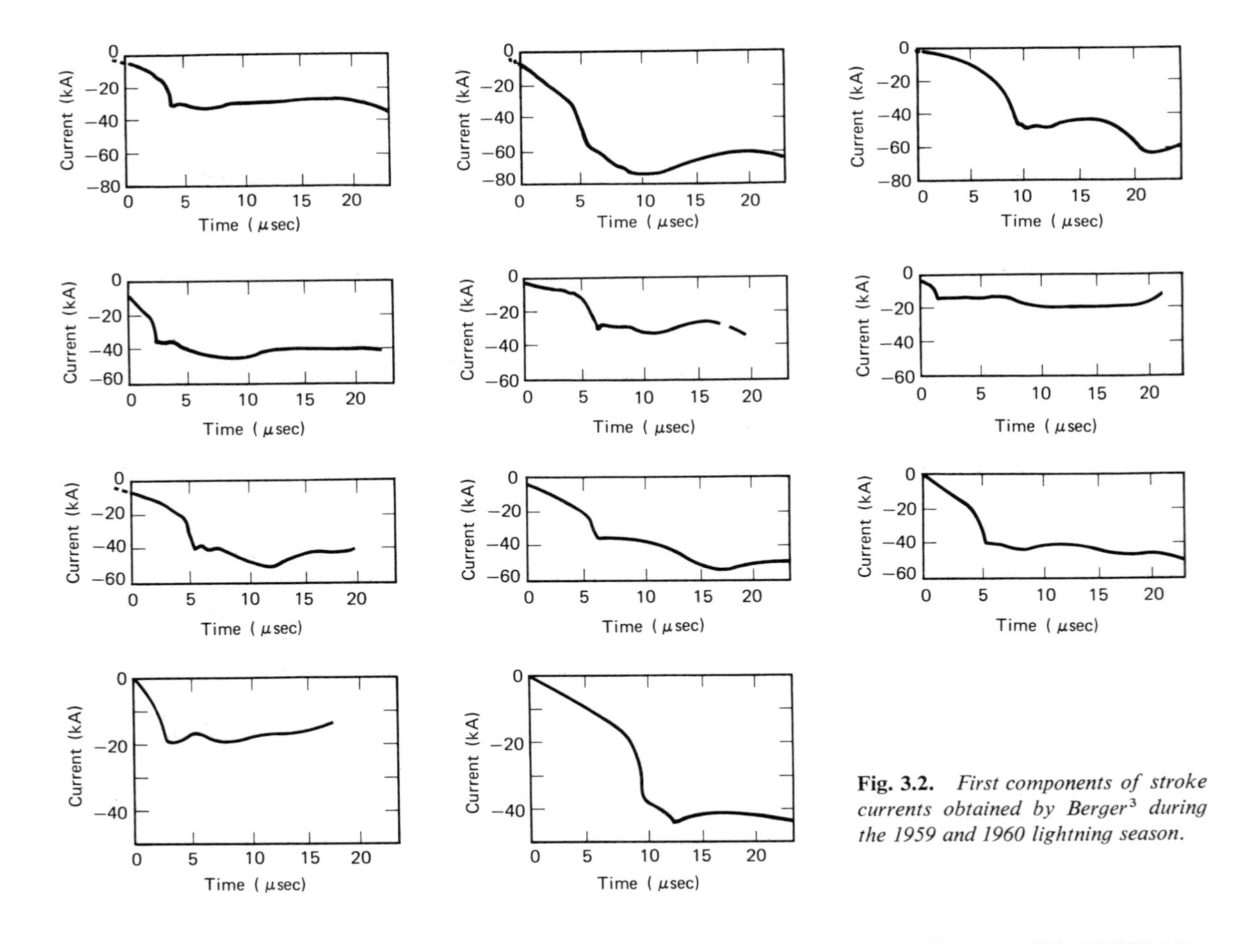

Fig. 3.2. *First components of stroke currents obtained by Berger[3] during the 1959 and 1960 lightning season.*

The magnitude of currents in lightning flashes terminating on grounded structures is indicated in Fig. 3.3.

Stroke currents to massive grounded structures like skyscrapers are generally large, probably because the capacitance between the leader channel and the building is greater than that between the leader and its image in the earth.[6]

There are two main types of lightning-discharge currents: (a) impulsive discharge, which reaches a crest value of 10 or 20 kA in 1 to 2 μsec and has a duration to half-value of 100 to 1000 μsec, and (b) nonimpulsive discharge of 100 A or so, lasting from 10 to 100 msec. Between 80 and 90% of flashes to earth contain impulsive strokes, and 50 to 60% contain non-impulsive strokes. This indicates that flashes to ground often contain both types. The predominance of the impulsive type is attributed to the earth's being a low-impedance-current source.

Lightning strokes are commonly described by two figures, such as 1.5 × 40, which means a current rise time of 1.5 μsec and a 40-μsec period for the current to fall to one-half its crest value.

In North America about half the lightning discharges will have crest values

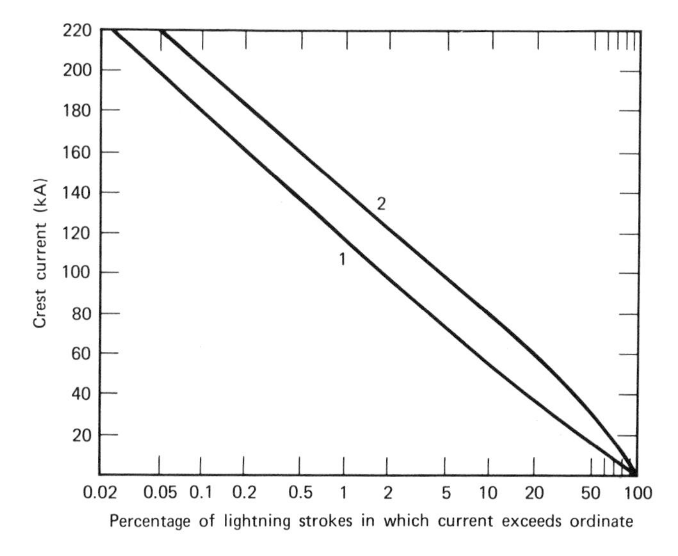

Fig. 3.3. *Distribution of crest currents in lightning strokes. Curve 1: currents in strokes to transmission-line ground structures; based on 4410 measurements, 2721 in the United States and 1689 in Europe. Curve 2: currents in strokes to buried structures, derived from curve 1. After Sunde.*[7]

exceeding 17,000 A, and an extreme value of 240,000 A will occur in about one stroke out of every 10,000.

3.7 PROBABILITY OF A LIGHTNING STROKE

As would be expected, the frequency of lightning flashes is proportional to the number of thunderstorm days. The following formula has been developed[6] for estimating the frequency of lightning flashes per thunderstorm day, taking into account the distance of the location from the equator.

$$N_E = (0.1 + 0.35 \sin \lambda)(0.40 \pm 0.20)$$

where N_E is the number of flashes to earth per thunderstorm day per square kilometer and λ is the geographical latitude. From this expression the risk of a particular structure's being struck by lightning can also be estimated. The area over which the structure can be expected to attract a lightning strike depends on the intensity of the stroke. The sketch in Fig. 3.4 shows the

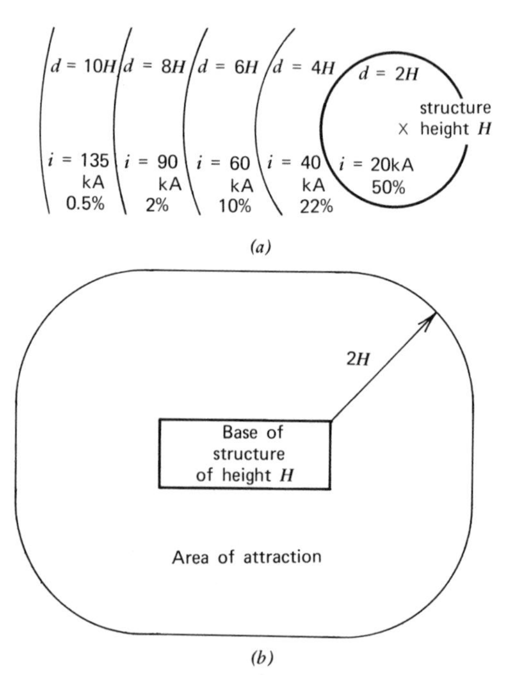

Fig. 3.4. *Attractive area of a conducting structure: (a) variation of attractive distance of with magnitude of lightning current i and frequency of its occurrence; (b) average area of attraction. After Golde.*[6]

radial distance to the edge of the attractive zone for a range of lightning-current values and indicates the relative frequency at which these current magnitudes occur. It will be seen that the radial distance varies from a value equal to twice the height of the structure, that is, $2H$, to a value of $10H$.

For estimating the probability of a strike let us use the value $2H$, which corresponds to the current magnitude (20 kA) of 50% of lightning flashes. Then if the length of the structure is L meters and its width is W meters, the total attractive area is

$$LW + 4H(L + W) + 4H^2\pi$$

For $L = 50$, $W = 30$, and $H = 30$ m, the attractive area is 0.022 km^2. If the structure is located at latitude 50°, for example, one can calculate from the expression for N_E that the frequency of lightning flashes to earth per thunderstorm day per square kilometer is approximately 0.01. The average number of thunderstorm days per year has been determined for a number of countries. Suppose the number is 30 for our example. Then there is the probability of one flash occurring per $1/0.30 = 3.3$ per square kilometer of area per year; thus the structure is likely to be struck once every 3.3/0.02 or 160 years. This sort of figure is of interest to those who wish to weigh the cost of installing a grounding system for a structure against purchasing insurance and to those who underwrite the insurance.

3.8 THE NUMBER OF THUNDERSTORM DAYS

In order to gather specific data on cloud-to-ground discharges in particular areas, various devices have been used to count the number and intensity of lightning strokes. One type of instrument adopted by the International Radio Consultative Committee (CCIR) is sensitive over the frequency range 2 to 50 kHz, and a second type adopted by the Conference Internationale des Grandes Réseaux Electriques (CIGRE) has a sensitive range of 120 Hz to 2 kHz. The CCIR is primarily interested in the assessment of interference to radio communication, and the CIGRE is concerned with the lightning hazard to power-transmission lines and systems. It has been established that the higher frequency radiation is dominant in intracloud discharges, the lower frequencies emanating mainly from a lightning return stroke to ground.

A new development in lightning-counter design[8] combines the output from a CIGRE type of counter, with a noise measurement in the 10-MHz range, to respond to ground flashes only or to discriminate between ground

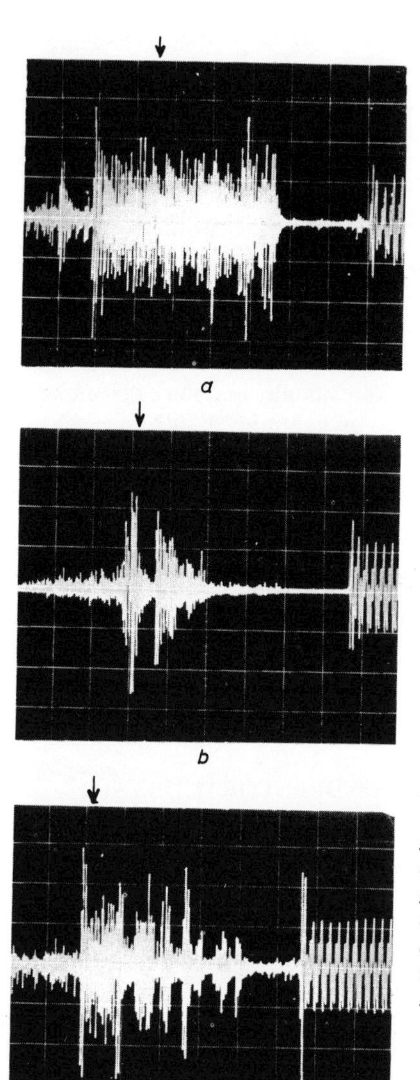

Fig. 3.5. *Noise patterns received at 10.01 MHz from thunderstorms at distances of less than 15 km: (a) typical cloud flash; (b) typical ground flash; (c) flash starting with a cloud-discharge phenomenon followed by several ground strokes. The triggering point of the extra-low-frequency receiver is indicated in each case by a small arrow, determined by the 50-Hz reference trace, which is introduced 540 msec after the counter has been triggered. Vertical axis: median value at receiver input ~ 2 mV peak to peak. Horizontal (time) scale: 100 msec per division. From Kreielsheimer and Lodge-Osborn.*[8]

and cloud flashes. The instrument takes advantage of a gap in high-frequency noise, which follows a ground stroke. The noise patterns of cloud and ground flashes are illustrated in Fig. 3.5. The block diagram of this "universal" counter[8] is shown in Fig. 3.6. It counts the total number of flashes within its range of sensitivity and discriminates between cloud and ground flashes.

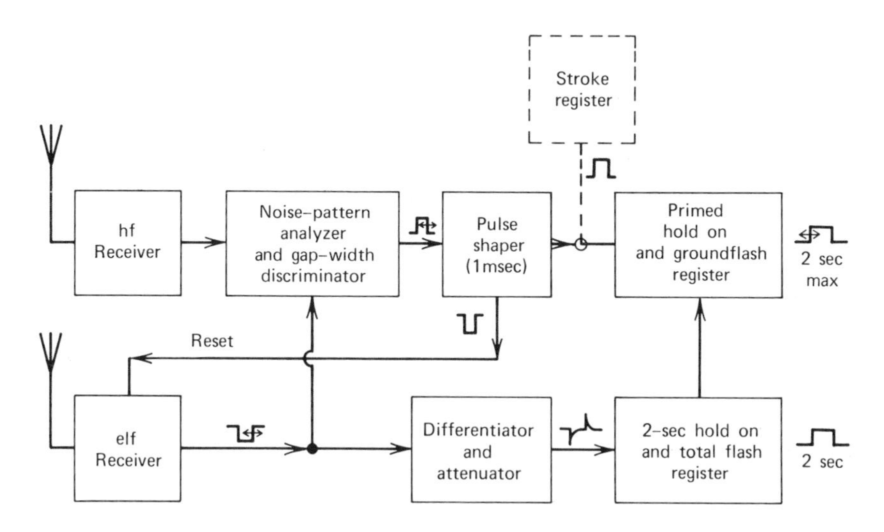

Fig. 3.6. *Block diagram of universal counter. After Kreielsheimer and Lodge-Osborn.*[8]

The data from lightning-flash counters in the past have been used to prepare isoceraunic maps like the examples in Figs. 3.7 (Great Britain) and 3.8 (United States), and the world map of Fig. 3.9.

3.9 THE FIRE RISK OF LIGHTNING

It has been indicated by laboratory experiments that a relatively long "tail" on the lightning current wave is required to cause a fire. For example, a current surge of 10 kA lasting for 10 μsec will not set fire to wood. Observations by Fuguay et al.[11] in 1967 determined that seven lightning discharges that caused forest fires each contained at least one long continuing tail, exceeding 40 msec in duration. Of 856 cloud-to-ground lightning discharges observed at the time, about one-half possessed a long continuing current tail.

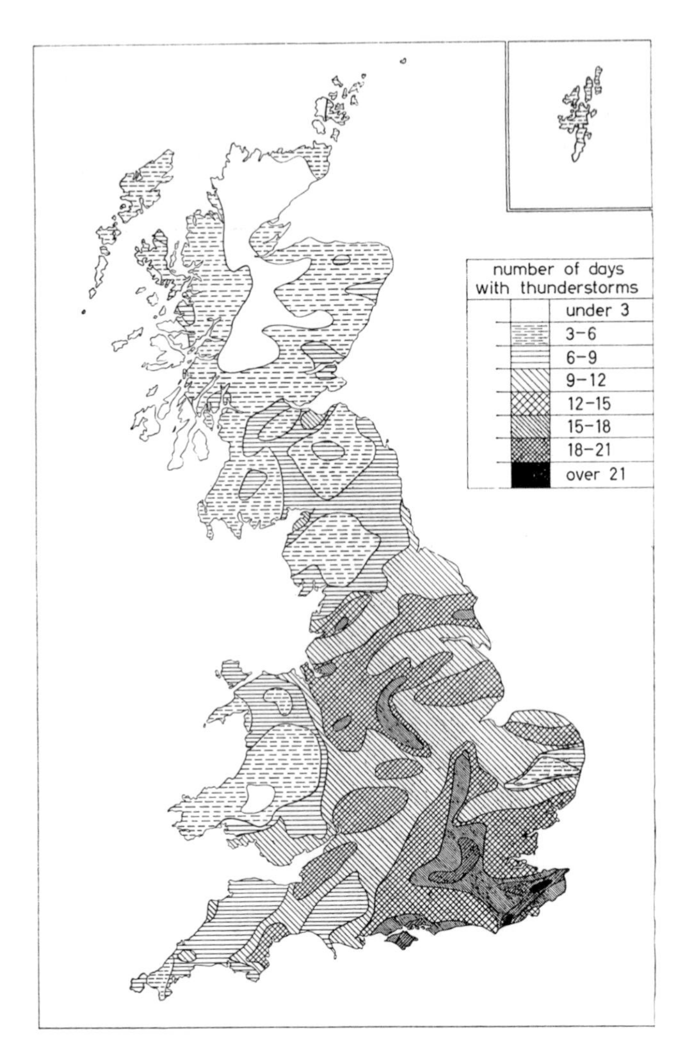

Fig. 3.7. *Average number of thunderstorm days in Great Britain during 1955–1959. This map has been prepared from reports received from over 1000 voluntary observers. From Golde.*[9]

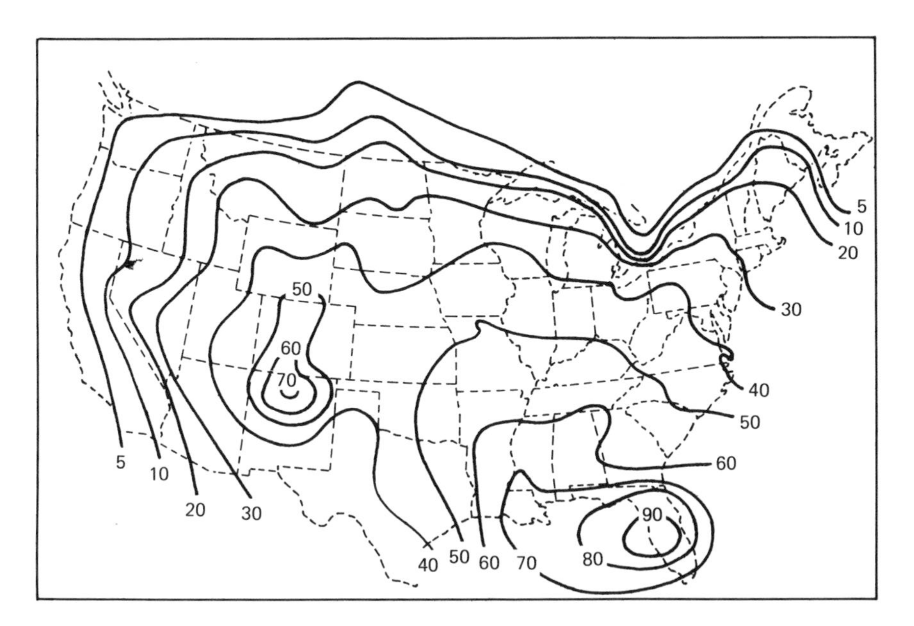

Fig. 3.8. *Map showing the average number of thunderstorm days per year in the United States.*

REFERENCES

1. B. J. Mason, "Mechanism of the Lightning Flash," *Electronics and Power,* May 1966.
2. B. B. Phillips, "Convected Charge in Thunderstorms," *Mon. Weather Rev.,* **95,** No. 12, December 1967.
3. K. Berger, "Novel Observations on Lightning Discharges," *J. Franklin Inst.,* **283,** No. 6, June 1967.
4. M. A. Uman, *Lightning,* McGraw-Hill, New York, 1969.
5. D. J. Malan, "Physics of the Thunderstorm Electric Circuit," *J. Franklin Inst.,* **283,** No. 6, June 1967.
6. R. H. Golde, "Protection of Structures Against Lightning," *Proc. IEE,* **115,** No. 10, October 1, 1968.
7. E. D. Sunde, *Bell System Tech. Journal,* **24,** April 1945.
8. K. S. Kreielsheimer and D. Lodge-Osborn, "New Development in Lightning-Counter Design," *Proc. IEE,* **118,** No. 1, January 1971.
9. R. H. Golde, "Thunderstorms," *JIEE,* May 1962.
10. A. B. Wood, "Lightning and the Transmission Engineer," *Electronics and Power,* June 1969.
11. D. M. Fuguay et al., "Characteristics of Seven Lightning Discharges That Caused Forest Fires," *J. Geophys. Res.,* **72,** 6371 (1967).

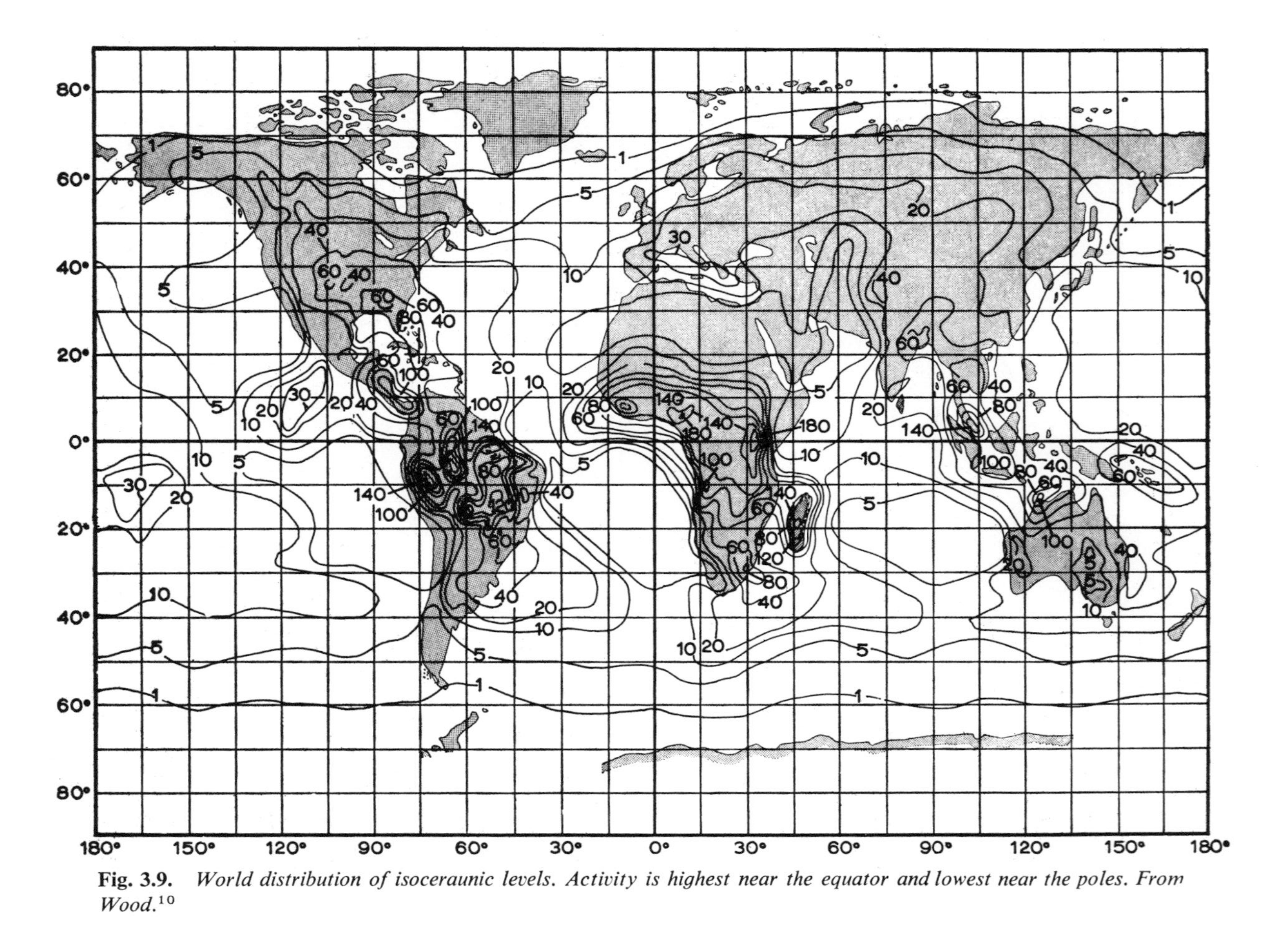

Fig. 3.9. *World distribution of isoceraunic levels. Activity is highest near the equator and lowest near the poles. From Wood.*[10]

CHAPTER 4

The Earth as a Discharge Terminal

4.1 SOIL-RESISTIVITY FACTORS

The earth is, of course, the terminal for lightning discharges, and when charge is suddenly transferred to earth or to a grounded object, it seeks to spread outward until neutralized by the ambient charge level of the whole earth mass. So the earth is the medium that dissipates electrical energy, and protection principles are intended to channel this energy so that it will be least harmful to property and life. The capability of the earth to accept the energy depends on the resistivity of the soil at the particular location and, in the case of structures, the effectiveness of the electrical connection to the earth.

The resistivity of the soil is a variable quantity that can be determined accurately only by measurements. The principal characteristics affecting soil resistivity are the following:

1. Soil type.
2. Concentration and composition of dissolved salts.
3. Moisture content.
4. Temperature.
5. Texture or grain size.
6. Compactness.

The resistance characteristics of soils and rock formations are shown in Fig. 4.1.

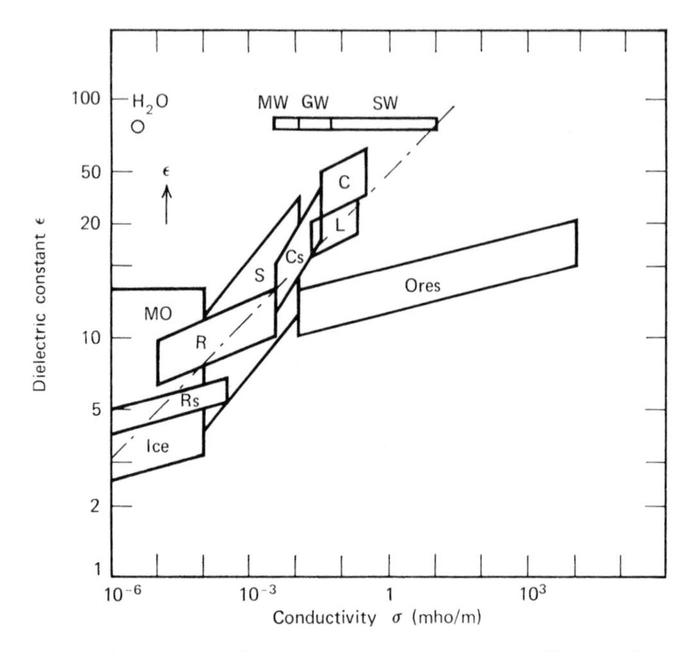

Fig. 4.1. *Conductivities and relative dielectric constants for soil types having natural moisture content and for water. Abbreviations: MW—mountain water; GW—groundwater; SW—seawater; C—clays; L—loam, marl, sandy clay; S—sand, gravels, sandstone, gypsum, limestone; Cs—clayey sand, shales; MO—mineral oil; R—crystalline rocks; Rs—rock salt, anhydrite. After Hanle.*[1]

The moisture content is an important variable in soil resistivity, not only by itself but also because it dissolves the various salts that occur in some soils. Therefore soil-resistivity measurements for ground systems should preferably be made under dry conditions so as to indicate the worst condition. There is a marked difference when the water is frozen. Frozen soil and ice have a comparatively high resistance. Fortunately there is less thundercloud formation during the winter season in the temperate and northern zones. Maps indicating the resistivity (or conductivity) of geographical regions are available for some countries, but this information is only of general interest to indicate what conditions can be expected for a large system like a power-transmission line.

The resistivity of the soil ρ or its reciprocal, the conductivity σ, will appear many times in dealing with grounding systems. The resistivity is defined as the resistance between opposite sides of a cube of unit dimensions. The unit ohm-meter (Ω-m) refers to the resistivity of soil (or other matter) calculated

TABLE 4.1 CONVERSION TABLE OF RESISTIVITY UNITS[a]

Unit	Ω–mm^2/m	Ω–cm	Ω–in.	Ω–m[b]	Ω/100 ft	Ω–km	Ω–circular mil/ft
Ω–mm^2/m	1	10^{-4}	3.9370×10^{-5}	10^{-6}	3.2808×10^{-8}	10^{-9}	6.0153×10^{2}
Ω–cm	10^{4}	1	0.3937	10^{-2}	3.2808×10^{-4}	10^{-5}	6.0153×10^{6}
Ω–in.	2.5400×10^{4}	2.5400	1	0.0254	8.3333×10^{-4}	2.5400×10^{-5}	1.5279×10^{7}
Ω–m[b]	10^{6}	10^{2}	39.37008	1	3.2808×10^{-2}	10^{-3}	6.0153×10^{8}
Ω/100ft	3.0480×10^{7}	3.0480×10^{3}	1.2000×10^{3}	30.48009	1	0.03048	1.8334×10^{10}
Ω–km	10^{9}	10^{5}	3.9370×10^{4}	10^{3}	32.8083	1	6.0152×10^{11}
Ω–circular mil/ft.	1.6624×10^{-3}	1.6624×10^{-7}	6.5450×10^{-8}	1.6624×10^{-9}	5.4542×10^{-11}	1.6624×10^{-12}	1

[a] All values are in absolute units.
[b] Système Internationale (SI) unit.

or measured between two faces of a 1-m cube. The term is sometimes expressed as ohms per meter or meter-ohms. Other units that are convertible to or from ohm-meters may be used. Table 4.1 lists the conversion factors for changing from one unit to another. The ohm-meter is recognized as the SI (Système Internationale) unit.

4.2 IMPEDANCE CHARACTERISTICS

Although soil resistivity is the property of main concern in grounding systems, some mention should be made of the dielectric constant and permeability. The electromagnetic properties of the earth are referred to a 1-m cube of the material. The resistivity ρ is expressed in ohm-meters or ohms per meter, and the conductivity σ, in mhos per meter. (They may also be expressed per centimeter (cube) rather than per meter.) The dielectric constant ε and permeability μ are expressed as numbers relative to unity.

The resistivity and effective dielectric constant may vary considerably within a short distance due partly to varying layer thicknesses in stratified earth, and resistivity in a direction parallel to stratified layers will often be greater than that in a perpendicular direction. Specific pockets in the earth may differ in resistivity from the general value. Deposits of mineral ores or graphites would have a relatively high conductivity. As a matter of fact, surveying for mineral ores is sometimes done with high-frequency radio transmission and reception. Most pure mineral rocks, however, have a high resistivity of 10^5 or 10^6 Ω-m and a dielectric constant of 2 or 3. The resistivity of rock deposits can vary with the moisture content and with the amount of salts dissolved therein. There is a relationship between the dielectric constant and conductivity that one source[1] expresses as $\varepsilon \approx 50\sigma^{1/5}$ (σ in mhos per meter) and further states that due to moisture content the dielectric constant is modified by a factor of $2.5 \times 0.78 \times$ water content (%).

Moisture content in earth can vary from 5% under dry conditions to 40% just after a rain. Values of 10 to 30% are common. In Fig. 4.2 the effects of water content on the resistivity of three soil types are shown. It is apparent that, when the water content becomes high enough, about 20%, the resistivities of the three soils approach the same value.

The impedance of the earth's surface layer is nearly independent of frequency under about 30 MHz, but the depth of penetration varies with frequency f (in hertz) as $(\pi f \sigma \mu)^{-1/2}$. The depth of penetration also depends on the relative conductivities of the earth layers below.

The degree of penetration with frequency and average conductivity is illustrated in Table 4.2, which indicates the penetration in meters to the depth where the current has dropped to 37% of its initial value.

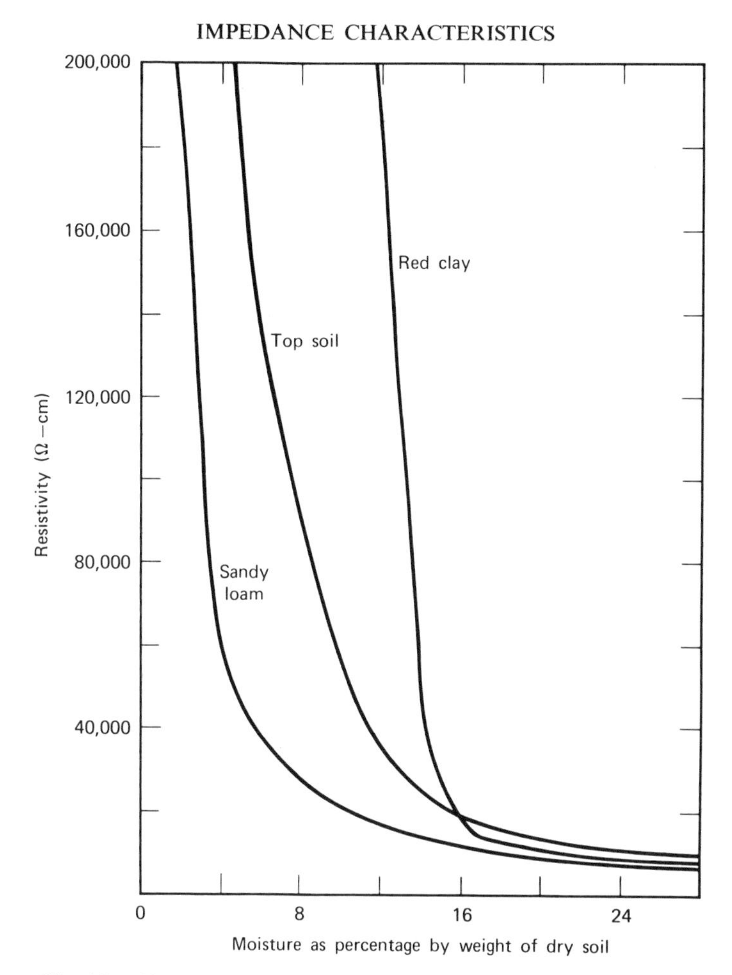

Fig. 4.2. *Variation in soil resistivity with moisture content. After Tagg.*[2]

TABLE 4.2 PENETRATION AS A FUNCTION OF FREQUENCY AND CONDUCTIVITY

	Penetration (m)			
Frequency	Conductivity		(mmho/m)	
(KHz)	1	3	10	30
100	46	28	17	10
500	20	11	7	4.5
1000	15	8.5	5	3
1500	12	7.5	4.5	2.5

In layered earth the average dielectric constant may be quite different from the value for a particular layer. The relative dielectric constant varies from 1 to 80, but ordinarily it is not a significant factor in grounding systems for lightning. For example, the relative dielectric constant of ice is $2\varepsilon_0$ and that of saltwater is $80\varepsilon_0$, where ε_0 is the dielectric constant of air. The dielectric strength, however, is of importance because it determines the extent to which the soil surface will ionize radially outward from a point struck by lightning.

4.3 MEASUREMENT OF EARTH RESISTIVITY

In the following sections on earth-resistivity measurements and the calculation and measurement of ground electrode systems, most of the data and curves were obtained from Tagg.[2]

The most generally used method of measuring earth resistivity is the four-electrode method. Referring to Fig. 4.3, we see that four metal rods whose diameter is small relative to their length and which are exposed only at the end are driven into the ground. A known current is passed from electrode 1 to electrode 4, and the potential drop is measured across electrodes 2 and 3. The current density flowing into the earth from electrode 1, spread over a spherical surface of radius r is $1/4\pi r^2$ times the value of entering current. For unit current the resulting potential gradient is the current density times the earth resistivity, or

$$-de/dr = \frac{\rho}{4\pi r^2}$$

where ρ is the resistivity and e is the potential at the radial distance r.

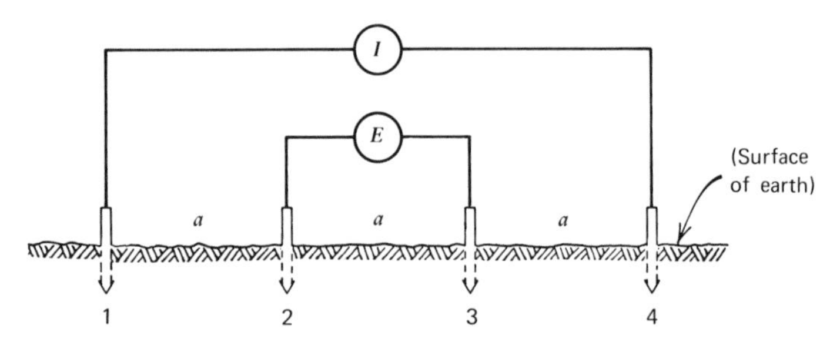

Fig. 4.3. *Four buried rods in line.*

The difference in potential between two points at distances r_1 and r_2 from electrode 1 would be the integral of the potential gradient between the two radii, that is,

$$e_1 - e_2 = \frac{\rho}{4\pi} \int_{r_1}^{r_2} \frac{dr}{r^2} = \frac{\rho}{4\pi} \left(\frac{1}{r_1} - \frac{1}{r_2} \right)$$

If the electrodes are spaced a distance a apart, as in Fig. 4.3, so that r_1 can be called a and r_2 is $2a$, then the potential difference between electrodes 2 and 3 is $\rho/2\pi a$; and if the current value is made unity, the resistance R is $\rho/2\pi a$ or ρ is $2\pi aR$. This derivation can be generalized, but it has been simplified here to electrodes placed in a straight line and equally spaced. Measurements are usually made with this arrangement of electrodes, and instruments are available to read the value of R directly, from which ρ is obtainable by simple calculation. The value of resistivity obtained is the equivalent value over the distance spanned by the electrodes. Variations in the electrode arrangement can be made[2] to evaluate inhomogeneities in the earth, using a suitable variation in the resistivity formula.

4.4 EARTH ELECTRODE—BURIED HEMISPHERE

In grounding problems it is sometimes permissible to regard the base of a structure in the earth as an equivalent hemisphere. This form of electrode is also a simple one to begin with in calculating electrode resistance. If we consider a buried hemispherical electrode as sketched in Fig. 4.4, the elemental resistance across the thickness dx of a hemispherical shell of the earth at radius x is

$$dR = \frac{\rho\, dx}{2\pi x^2}$$

The whole resistance between the surface of the electrode of radius r and some electrode at distance r_1 is

$$R \int_{r}^{r_1} \frac{\rho\, dx}{2\pi x^2} = \frac{\rho}{2\pi} \left(\frac{1}{r} - \frac{1}{r_1} \right)$$

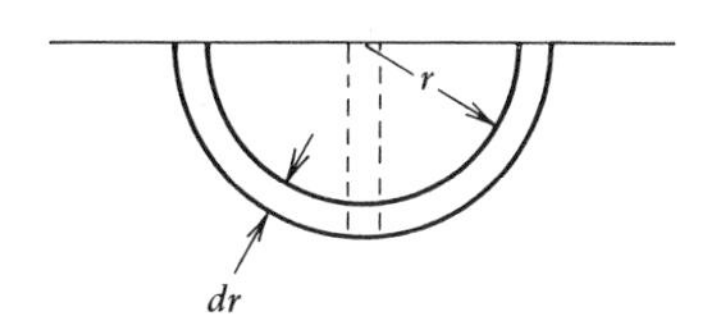

Fig. 4.4. *Equivalent hemispherical surface of buried rod electrode.*

If r_1 is made infinity, the total resistance of the electrode is $R = \rho/2\pi r$. The manner in which the resistance varies with distance from the electrode is sketched in Fig. 4.5. It will be noticed that the curve flattens out after rising rather steeply. This characteristic leads to a convenient method of measuring the resistance to earth of ground electrodes. It is called the fall-of-potential method.

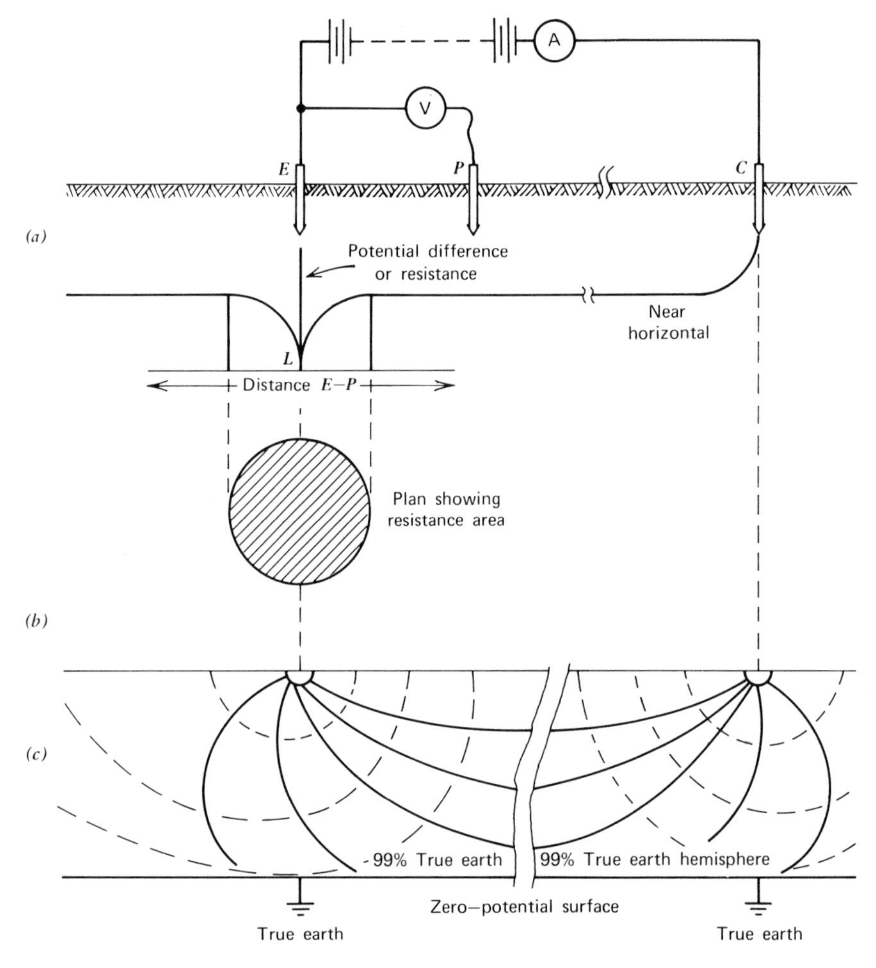

Fig. 4.5. *Variation of resistance with (a) distance between (b) equivalent hemispherical electrodes; and (c) current flow (solid lines) and equipotential lines (broken lines).*

4.5 RESISTANCE OF EARTH ELECTRODES

The most reliable method of measuring the resistance to earth of a buried electrode is the fall-of-potential method. Figure 4.6 shows an arrangement of three electrodes and a corresponding resistance curve. Let E be the electrode whose resistance to earth is required and let P and C be auxiliary rods driven into the earth. A known value of current I is circulated between C and E, and the voltage drop V between E and P is measured. Providing C is at an adequate distance from E, it has no appreciable effect on the resis-

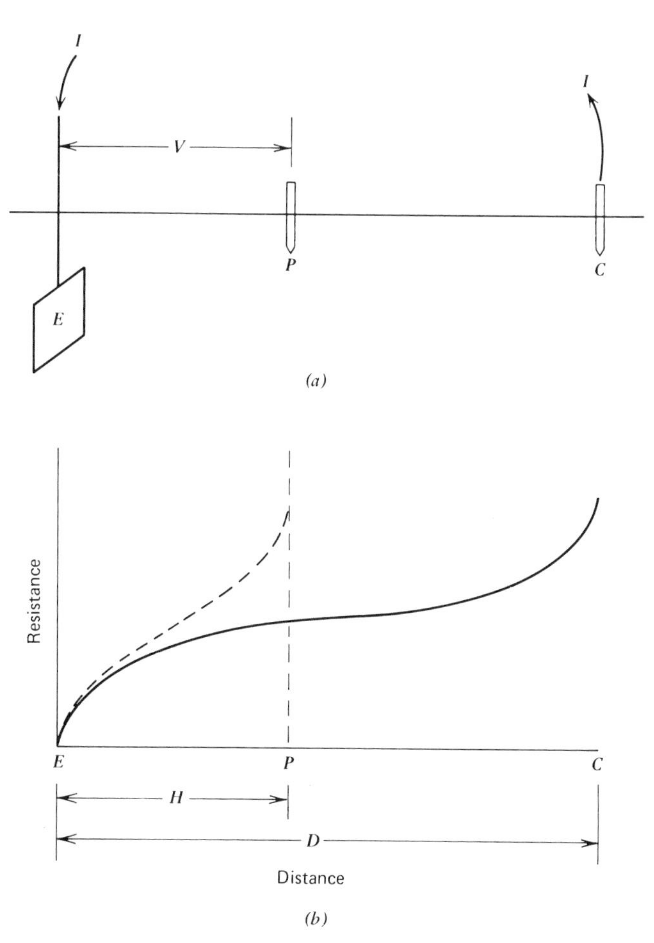

Fig. 4.6. *Variation in the earth resistivity of a buried electrode to earth with the spacing of the potential-measuring electrode (the fall-of-potential principle).*

tance value of E, but is simply a current source. If a negligible portion of the current I flows in rod P, it too will have a negligible effect on the resistance value of E. If in making this measurement the current electrode C were placed close to E and the potential electrode were placed at several points in between, a resistance curve such as that shown by the broken curve, with a relatively steep slope, would be obtained. However, if the spacing of electrode C is extended, a portion of the resistance curve will be nearly horizontal, indicating that the value would not change much even if the spacing were infinity. The up slope of the curve at electrode C shows the effect of its resistance, but, as already stated, the V/I measurements between E and P are independent of it. The resistance of electrode E to earth is V/I. The required spacing between E and C for a measurement with 98% accuracy is about 30 times the depth of the electrode E, and the optimum location for the potential electrode P is 0.62 of the distance from E to C. The derivation of this value is given in Appendix 4.A.

The curves of Fig. 4.7 show the required separation of current and poten-

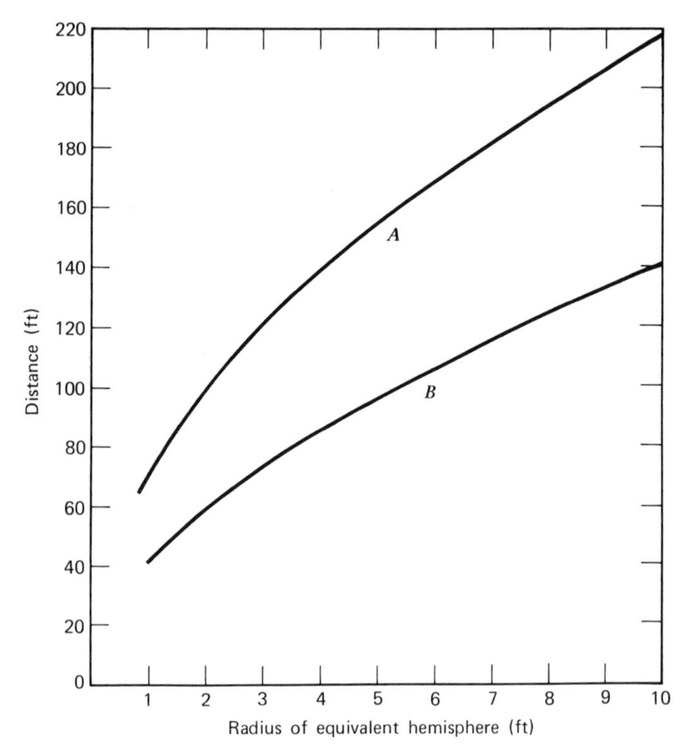

Fig. 4.7. *Spacing of auxiliary electrodes for 2% error. Curve A: distance to current electrode; curve B: distance to potential electrode. After Tagg.*[2]

tial electrodes from an electrode of known equivalent radius whose resistance is to be measured.

The equivalent radius is the radius of a hemisphere equivalent to an electrode or system of electrodes that might be rods, wires, plates, or other shapes. These will be considered in Section 4.13, after the earth resistance of single rods and other electrodes is derived.

4.6 SINGLE DRIVEN ROD—RESISTANCE TO EARTH AND EQUIVALENT RADIUS

The most familiar ground electrode is the driven rod or pipe, commonly used for low-voltage electric power distribution systems and telephone systems. Higher voltage substations or communications installations would use a number of ground rods properly spaced and connected in parallel, and more sophisticated systems employ both rods and horizontal conductors.

To find the equivalent hemisphere to a rod, for later convenience in calculations, it is necessary to find a formula for its resistance to earth. One such formula is derived in Fig. 4.8 and Appendix 4.B; it states that the resistance of the rod to earth is

$$R = \frac{\rho}{2\pi l}\left(\log_e \frac{4l}{a} - 1\right)\,\Omega$$

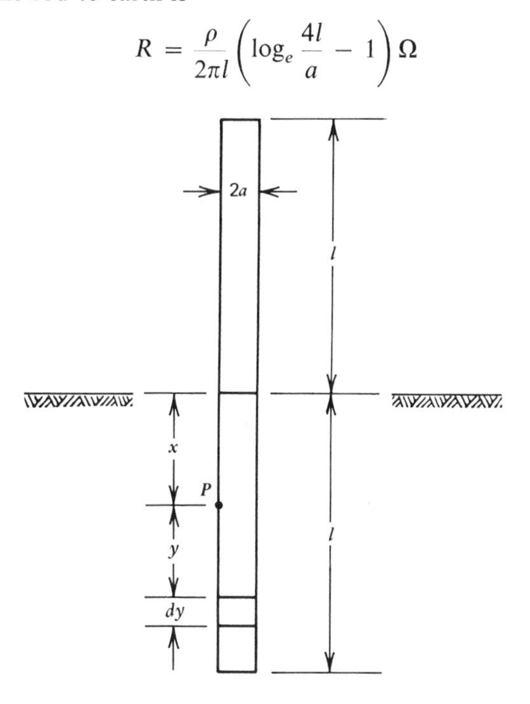

Fig. 4.8 *Single ground rod—derivation of resistance.*

and using this expression the radius of its equivalent hemisphere is found to be

$$r = \frac{l}{\left(\log_e \dfrac{8l}{d} - 1\right)},$$

where l is the buried length of rod, a is the radius of the rod, and d is $2a$, the diameter of the rod, all in the same units, and consistent with the unit of length used in ρ, the resistivity.

Using other units the resistance can be expressed as

$$R = \frac{\rho}{1.915l}\left(\log_e \frac{48l}{d}\right) \Omega$$

where ρ is in ohm-meters, l is in feet, and d is in inches. If rods 10 ft long and 0.75 in. in diameter are used, $R = 0.337\rho$, and from this a rough estimate of the net resistance of several rods adequately spaced can be obtained by dividing by the number of rods. An empirical formula[3] recommended for industrial grounds is $0.75R \langle n \rangle 0.25R$ for n rods of the size given above.

A list of equivalent radii of various rod sizes is given in Table 4.3.

TABLE 4.3 EQUIVALENT RADII

Length of Rod or Pipe (ft)	Radius (ft) of Equivalent Hemisphere for Rod or Pipe Radius of		
	0.5 in.	1 in.	2 in.
3	0.56	0.64	0.76
4	0.71	0.81	0.94
5	0.85	0.97	1.12
6	0.99	1.12	1.29
7	1.13	1.27	1.45
8	1.26	1.42	1.62
9	1.41	1.56	1.77
10	1.52	1.70	1.93

It will be apparent that the effect of increasing the rod radius is small. This is illustrated further by the chart of Fig. 4.9, which shows the comparative effects on ground rod resistance by varying the length of a 0.75-in.-diameter rod and by varying the diameter of a 10-ft rod.

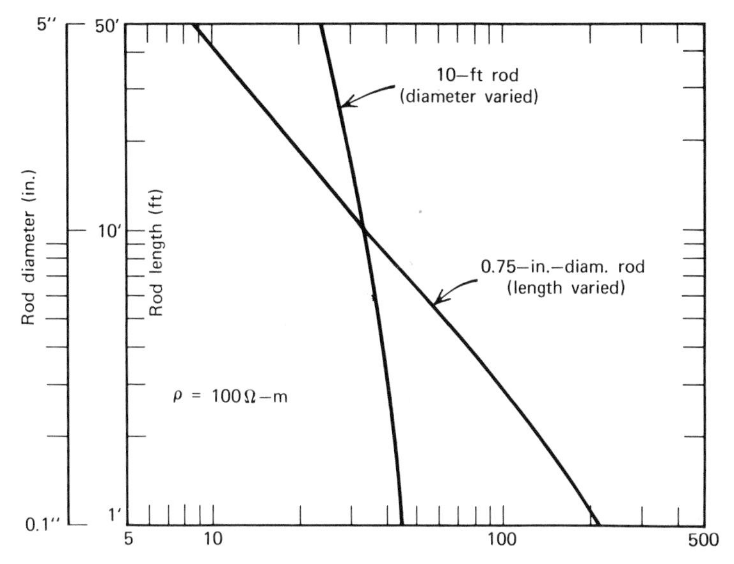

Fig. 4.9. *Effect of rod length and diameter on ground rod resistance. After Mukhedkar and Demers.*[3]

Figure 4.6 shows the variation of resistance with distance from an earth electrode. As the total current flowing into the earth must remain the same at any distance, this curve must also represent the potential difference between the electrode and any point; its slope is the potential gradient.

From Section 4.4 it will be recalled that the resistance at a distance r from an electrode of radius r is

$$R = \frac{\rho}{2\pi}\left(\frac{1}{r} - \frac{1}{r_1}\right).$$

The fractional reduction of resistance due to the distance in the earth from r to r_1 is

$$\frac{1/r - 1/r_1}{1/r} = 1 - \frac{r}{r_1}.$$

Therefore, to find the distance at which an electrode reaches say 90% of its ultimate (minimum) value, one would equate $1 - r/r_1$ to 0.90. Let us apply this to a 6-ft rod of 1 in. diameter whose equivalent radius is 1.12 ft: $1 - 1.12/r_1 = 0.90$. Then $r_1 = 11.2$ ft is the distance where the rod would have 90% of its ultimate effectiveness as a ground electrode. To attain 95% of its ultimate value the distance r_1 would be 22.4 ft, but then, owing to the diminishing slope of the potential gradient, it would require a distance of 112 ft to reach 99% of its ultimate value. It will be evident that in a multiple-rod system there should be a substantial spacing between rods to avoid a significant duplication of their functions, but that it would be impractical to try to realize the ultimate resistance value of each rod.

4.7 MULTIPLE DRIVEN-ROD ELECTRODES

As indicated, elaborate ground systems require multiple electrodes; and the higher the earth resistivity, the greater will be the number of electrodes required. Exact methods of calculating the resultant resistance of a number of rods connected in parallel are lengthy, but fortunately are not warranted because the soil resistivity is nonuniform and cannot be determined to a high degree of accuracy.

A suitable approximate method is one that uses the equivalent hemispherical electrode for each rod. Assuming that each electrode carries the same charge, the method entails calculating the potential, then the capacity, and from that the resistance to earth.

The potential of a sphere of radius r and carrying charge Q is $V = Q/r$. Also the potential at some distance d (in air) is $V_d = Q/d$. For a hemisphere we could use half this value, but as the equivalent spherical shape is only used as a convenience in finding the ratio of the resistance of multiple rods to the resistance of a single rod, it is simpler to use the formula given here.

Multiple ground rods are usually spaced a distance at least equal to their depth so as to realize their full effect in lowering the ground-system resistance. This will be made evident by the calculations for rods in parallel.

4.8 TWO AND THREE GROUND RODS IN PARALLEL

If there are two rods in parallel spaced by distance d and the radius of each equivalent hemisphere is r, then the potential of either rod is

$$V = \frac{Q}{r} + \frac{Q}{d} = \frac{Q}{r}(1 + \alpha)$$

where Q is the charge on each electrode and α is the ratio r/d.

The total charge on both rods is $2Q$, and their combined capacity can be found from

$$\frac{1}{C} = \frac{V}{2Q} = \frac{1}{2r}(1 + \alpha)$$

Using this expression the combined resistance of the two rods can be found from the previously given expression, $R = \rho/2\pi C$, that is, $R = \rho/4\pi r(1 + \alpha)$.

As the resistance of a single rod would be $\rho/2\pi r$, the ratio of the combined resistance of the two rods in parallel to that of a single rod is

$$\text{resistance ratio, two rods in parallel} = \frac{\rho/4\pi r\,(1 + \alpha)}{\rho/2\pi r} = \frac{1 + \alpha}{d}$$

The lowest value that this ratio can approach is $\frac{1}{2}$ when d is very large compared to r. In practice it is found that, when the spacing is made equal to the length of the buried rod, there is little further reduction in the ratio as the spacing is increased. Values of this ratio for several rod lengths and spacings are given in Fig. 4.10. By a similar method it can be shown that the resistance of three rods in parallel (see Fig. 4.11) expressed as a ratio of the resistance of a single rod is

$$\frac{2\alpha + k}{2 + k}$$

This derivation is shown in Appendix 4.C. Values of this ratio for several rod lengths and spacings are shown in Fig. 4.12.

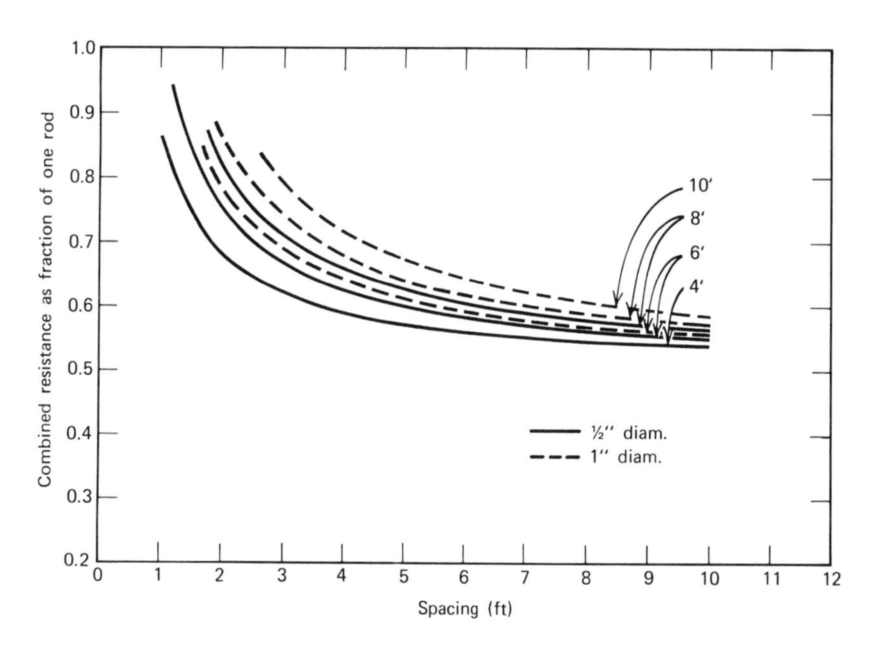

Fig. 4.10. *Combined resistance of two rods in parallel. After Tagg.*[2]

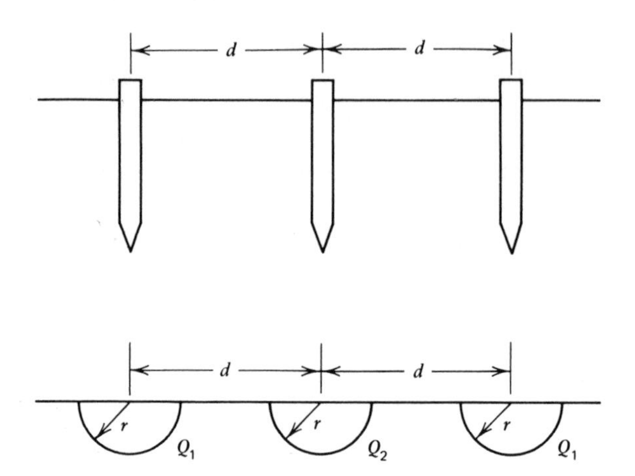

Fig. 4.11. *Three rods in parallel in a straight line.*

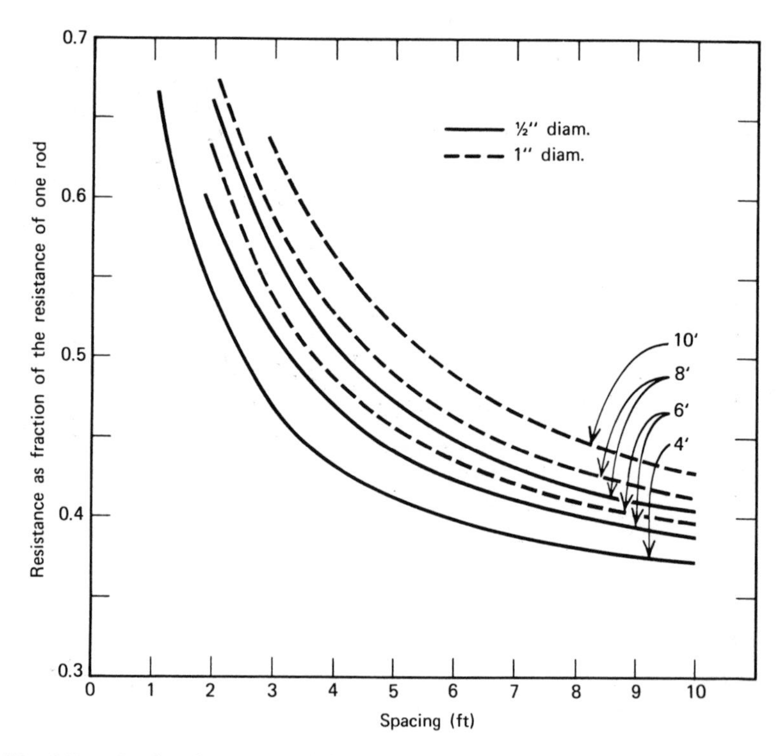

Fig. 4.12. *Combined resistance of three rods in parallel in a straight line. After Tagg.*[2]

4.9 RESISTANCE OF OTHER MULTIPLE-ROD ARRANGEMENTS

By similar methods we find that three rods placed at the corners of an equilateral triangle have a resistance $(1 + 2\alpha)/3$, where α is still r/d. This arrangement is nearly as effective as three rods in a line at the same spacing d.

A system of four rods in a straight line with spacing d, connected in parallel, will have a resistance ratio of $(12 + 16\alpha - 21\alpha^2)/(48 - 40\alpha)$ times the resistance of a single rod. The values of this ratio for various rod dimensions are given in Fig. 4.13.

4.10 RODS IN AN OPEN-SQUARE ARRANGEMENT

A compact arrangement of buried rods enclosing a structure to be grounded is a square, with a rod at each corner and others along the sides. If they are equally spaced, the rods are not necessarily all at the same potential, nor do they carry the same charge. However, it is permissible to assume the same charge on each rod. Then by calculating the potential of a corner rod and of a side rod, and taking the average potential, a suitable value for total capacity and resistance ratio can be obtained. If N is the total number rods in the square, then[2]

$$\frac{\text{resistance of } N \text{ rods in parallel}}{\text{resistance of one rod}} = \frac{1 - K\alpha}{N}$$

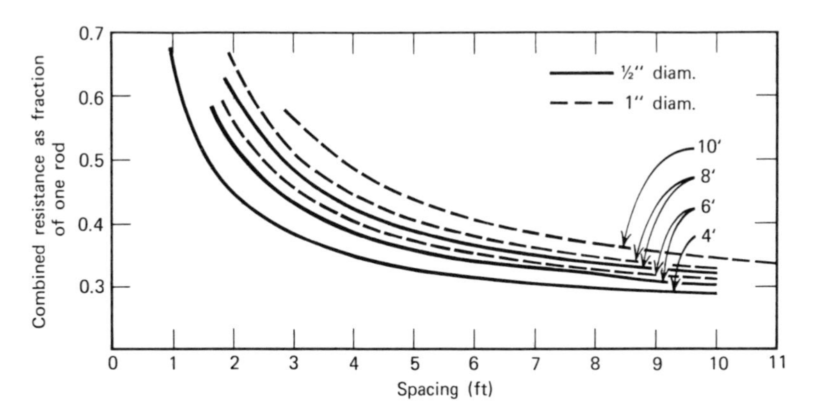

Fig. 4.13. *Combined resistance of four rods in parallel in a straight line. After Tagg.*[2]

where K depends on the number of rods used. Resistance-ratio curves for 8 rods and for 16 rods in an open-square form are given in Figs. 4.14 and 4.15, respectively. Values for 12 rods in a square lie about midway between these curves. It will be noted that the ratio improves at a greater rate, as the quantity and spacing increase, for short rods. However, a lower net resistance will be obtained with the longer rods.

More meaningful curves, which indicate the actual resistance for a given earth resistivity, are shown in Figs. 4.16 and 4.17. These are drawn for earth with a resistivity of 100 Ω-m and for 8- and 16-rod groups of several lengths, arranged in the form of an open square. The resistance of these systems for any other value of earth resistivity ρ, can be found by multiplying the value from the curve by $\rho(\Omega\text{-m})/100$.

If the area available for a grounding system is limited, it is well worthwhile to know the number of rods needed for an effective ground system, so that neither the available space nor copper rods are wasted. Figure 4.18 is an

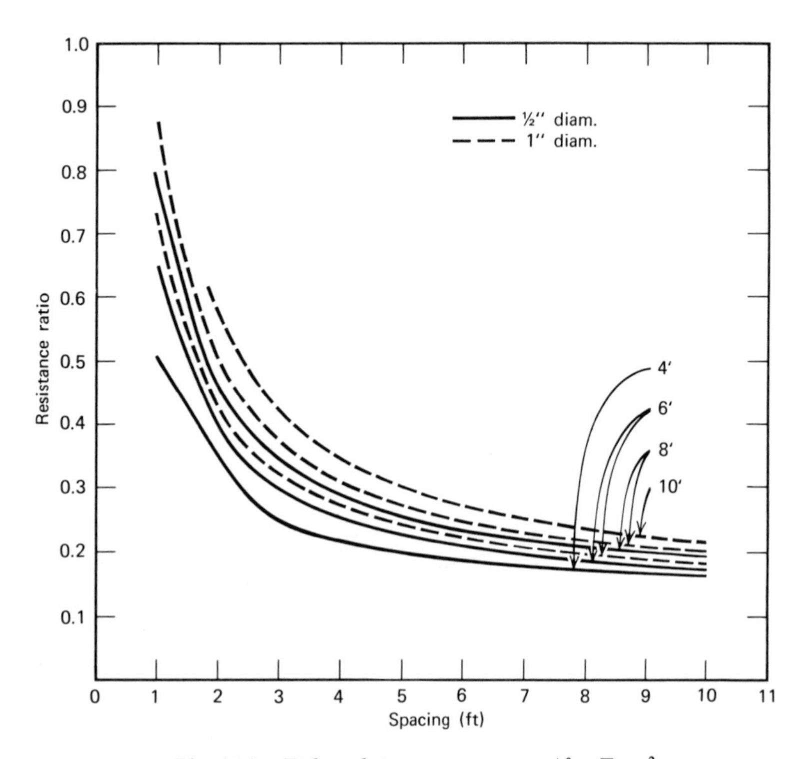

Fig. 4.14. *Eight rods in an open square. After Tagg.*[2]

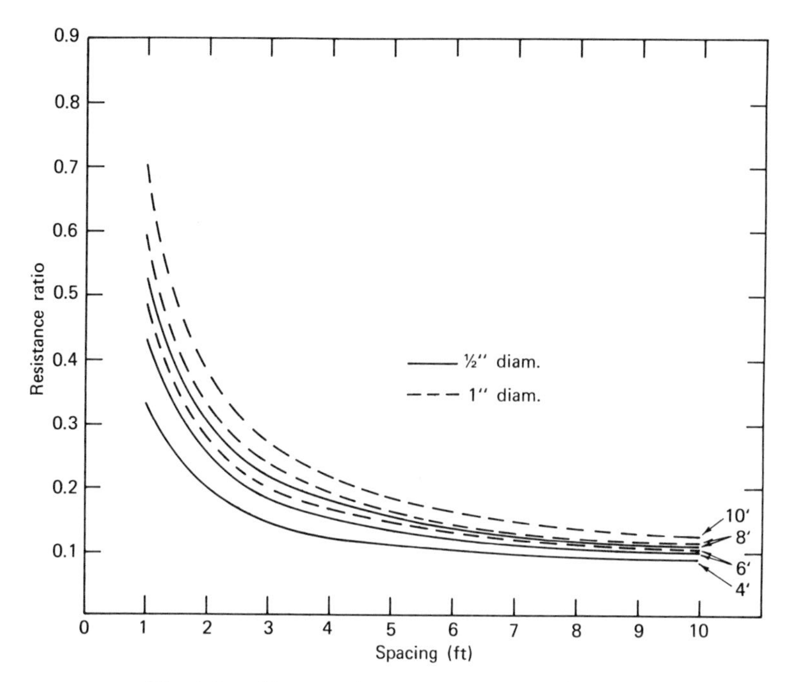

Fig. 4.15. *Sixteen rods in an open square. After Tagg.*[2]

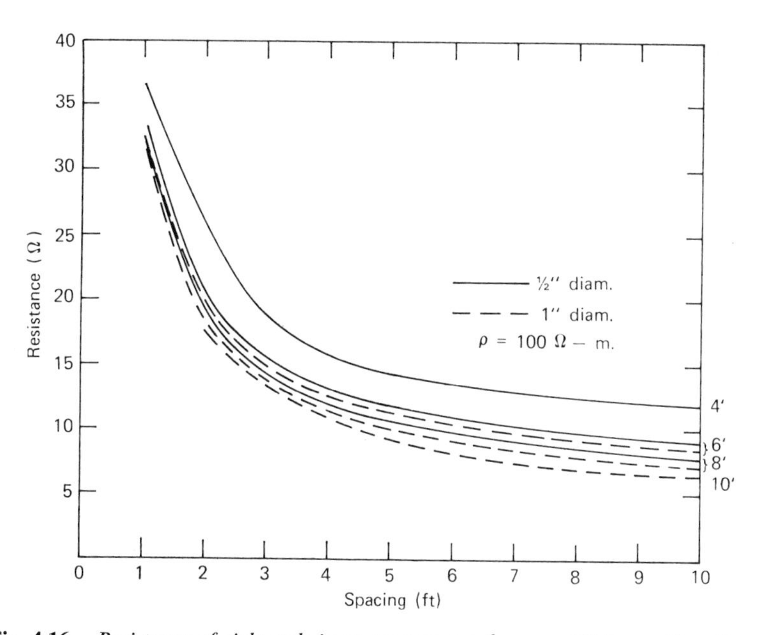

Fig. 4.16. *Resistance of eight rods in an open square for* $\rho = 100\ \Omega\text{-m}$. *After Tagg.*[2]

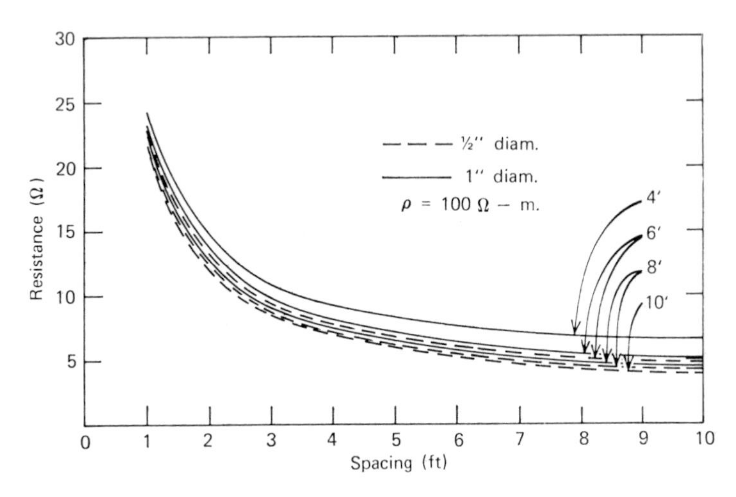

Fig. 4.17. *Resistance of 16 rods in an open square for $\rho = 100$ Ω-m. After Tagg.*[2]

example of such a curve, which was calculated for 8-ft rods arranged in an open square and $\rho = 100$ Ω-m. It indicates, for example, that in an area of 400 ft^2 15 rods or at most 20 rods are sufficient to approach the minimum resistance rather closely. The curve also indicates that, when a grounding area of say 2000 ft^2 is used, further reduction of only 0.5 Ω can be achieved by doubling that area. Accordingly an area of 1600 to 2000 ft^2 is sufficient, unless the earth resistivity is higher and/or it is necessary to achieve a ground-system resistance in the 3-Ω range.

The open-square model is applicable to ground systems placed in excavations for structures where the bedrock is close to the surface. In these circumstances it would be rather impractical to drill holes for ground rods in the surrounding area, and it might not be possible to bury horizontal conductors for connecting the rods together into one earth electrode system. A practical method is to lay conductors of No. 6 or No. 4 AWG wire vertically down each side of the excavation, spaced a distance equal to the depth of the hole, but not closer than 10 ft. At the bottom the tail ends may extend a few feet over the bottom of the excavation. These verticals should be connected together at top and bottom by horizontal conductors of the same size to form a "cage." If we assume that an excavation is 100 ft^2 and 10 ft deep, an estimate of its resistance to earth can be made if the resistivity of the soil material has been determined by measurement. From Fig. 4.18 it will be seen that the minimum resistance obtainable with 8-ft rods in a hollow

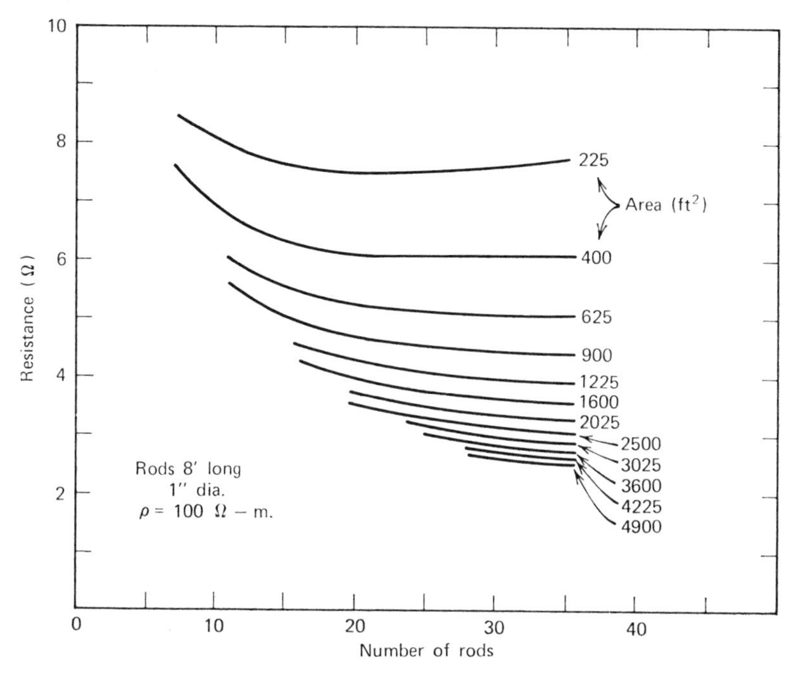

Fig. 4.18. *Resistance of rods (8 ft long, 1-in. diameter) in parallel in an open square of given area for $\rho = 100$ Ω-m. After Tagg.*[2]

square up to an area of 5000 ft^2 is 2.5 Ω for $\rho = 100$ Ω-m, or 10,000 Ω-cm, and this minimum is attained with approximately 30 rods. Accordingly for our example of a 100-ft^2 excavation, which would have 40 vertical conductors spaced 10 ft apart, a resistance value of 2.25 Ω for $\rho = 100$ Ω-m would be a close estimate. Let us assume that the resistivity of the actual rock material including backfill is 2000 instead of 100 Ω-m. Then the resistance of the system would be 20 × 2.25, or 45 Ω.

If additional rods are driven in the interior of a relatively small open square, spaced the same as the rods on the sides, it is found that very little reduction of resistance results. This is illustrated in Fig. 4.19, which also indicates the minimum resistance attainable for the rod size and earth resistivity cited.

A practical chart[3] for the resistance of grids of 10-ft ground rods in parallel where the rod spacing is greater than 20 ft is presented in Fig. 4.20, which shows the number of rods needed to attain a required resistance value for any value of earth resistivity.

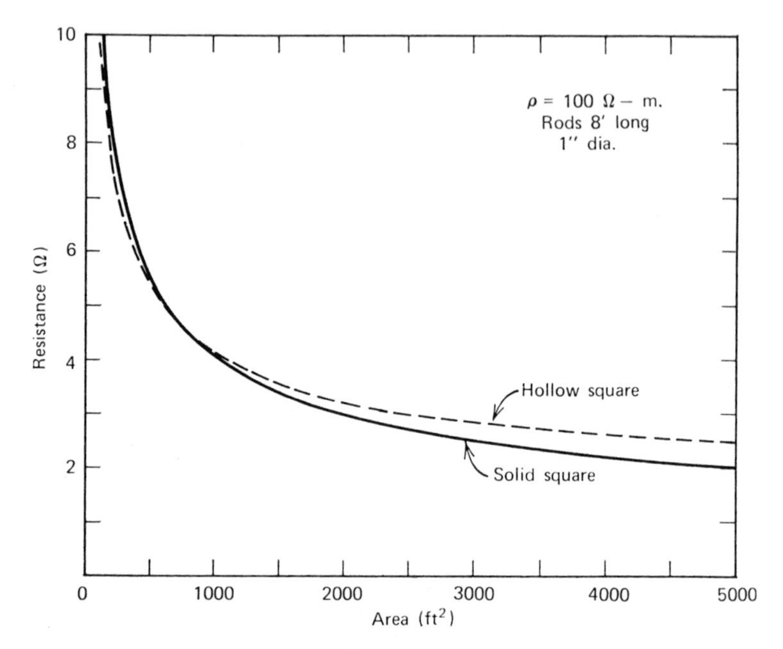

Fig. 4.19. *Minimum resistance obtainable with paralleled rods (8 ft long, 1-in. diameter) in a square of given area for $\rho = 100\ \Omega$-m. After Tagg.*[2]

4.11 MULTIPLE GROUND RODS ARRANGED IN A CIRCLE

Where convenient the driven rods can be positioned on the circumference of a circle, rather than on the sides of an open square. It is found that for a given circular area the ground resistance attainable will be about the same as for the same area in an open-square arrangement, assuming the same size and number of rods and the same earth resistivity. The resistance-ratio formula is[2]

$$\text{resistance of circle of rods} = \frac{1 + 0.5\alpha + \alpha \sum_{S=1}^{S=(n/2-1)} \operatorname{cosec} S\pi/n}{n}$$

where there are n rods in the circle of radius R feet, $\alpha = r/R$ (r being the equivalent spherical radius of the electrode), and s is the depth of the rods. The resistance to ground for groups of rods equally spaced on a circle of 20-ft radius expressed as a ratio of the resistance of a single rod is shown in

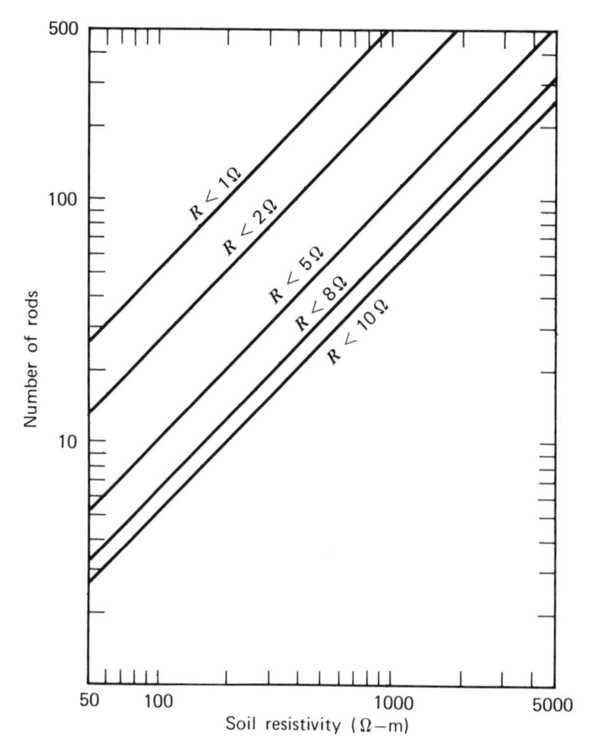

Fig. 4.20. *Number of rods (10 ft long, 0.75-in. diameter, spaced more than 20 ft apart) needed to attain a required resistance value R for any value of soil resistivity. After Mukhedkar and Demers.*[3]

Fig. 4.21. The curves are drawn for rods of several lengths and two values of diameter.

4.12 BURIED-WIRE GROUND ELECTRODES

The buried-wire ground conductor is applicable to high-voltage power-transmission lines and in areas where the surface soil is shallow. In the latter case an arrangement of wires fanning out radially from a central connection point is used. Furthermore, extensive buried radial wire systems are used at radio-transmitting stations, where they serve primarily as a ground return circuit for the high-frequency energy.

The method of calculating the resistance of a buried wire is similar to the procedure for a buried rod. The electrostatic capacity is first calculated, then

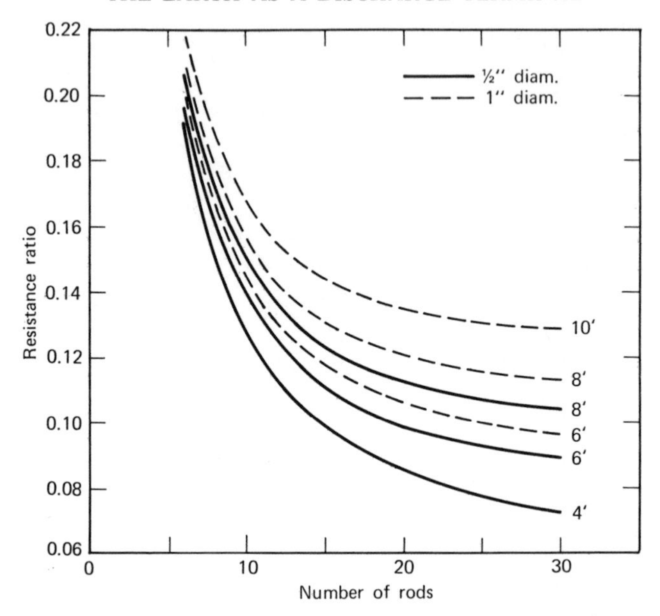

Fig. 4.21. *Rods in a circle of 20-ft radius. After Tagg.*[2]

a value for its potential, assuming a given charge, and from these two quantities an expression for the resistance to earth is attainable. The method is indicated in Appendix 4.D.

The expression for the resistance of a buried wire to earth takes the form

$$R = \frac{\rho}{4\pi l}\left[\log_e \frac{4l}{a} - 1 + \log_e \frac{2l + \sqrt{S^2 + 4l^2}}{S} + \frac{S}{2l} - \frac{\sqrt{S^2 + 4l^2}}{2l}\right]$$

where $2l$ is the length of buried wire, $S/2$ is the depth of the wire below the surface, and a is the radius of the wire.

Curves showing the value of R for various lengths and radii of wire are presented in Fig. 4.22.

Expressions for the resistance of several star arrangements of buried wire have been calculated[2] as an extension of the method for a single wire. All the formulas take the general form

$$R = \frac{\rho}{2\pi l_t \delta}\left(\log_e \frac{48l}{d} + f\frac{t}{l}\right)$$

where l_t is the total length of wire, l is the length of one arm of the system, d is the wire diameter in inches, and δ is the conversion factor from feet

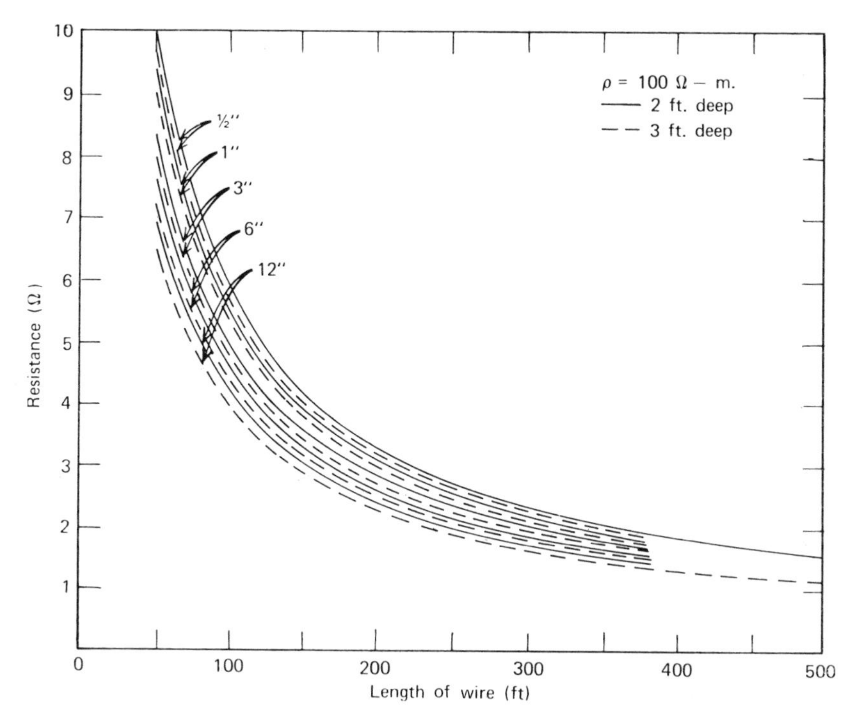

Fig. 4.22. *Resistance of horizontal wires of varying diameter (indicated by figures pointing to curves) buried 2 ft deep (solid curves) and 3 ft deep (broken curves) for $\rho = 100\ \Omega$-m. After Tagg.*[2]

to centimeters. The function f, (t/l), a depth factor, is different for each arrangement of wires, and a plot of this function for six wire patterns is shown in Fig. 4.23. Using the appropriate value from these curves and the nomogram of Fig. 4.24, one can find the resistance of the particular buried-wire system for $\rho = 100\ \Omega$-m. For other values of earth resistivity it is necessary to multiply by $\rho(\Omega\text{-m})/100$. The resistance values for several wire arrangements are shown in Fig. 4.25.

4.13 OTHER FORMS OF ELECTRODES

The buried wire may be placed in the form of circular ring. However, for a given length of wire its resistance to earth is about the same as a three-radial-star form, and the ring is usually less convenient.

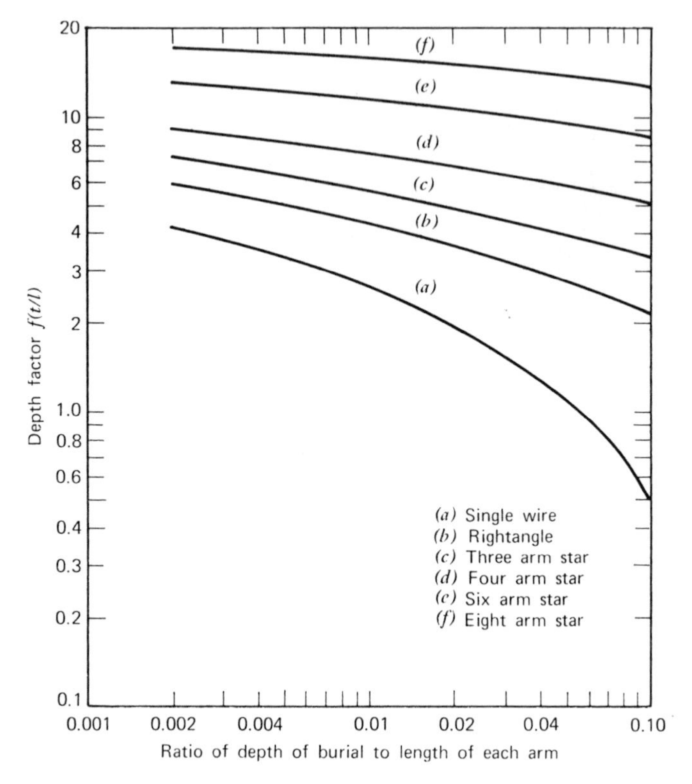

Fig. 4.23. *Depth factor for buried wires arranged in six patterns:* (*a*) *single wire;* (*b*) *right-angle pattern;* (*c*) *three-arm star;* (*d*) *four-arm star;* (*e*) *six-arm star;* (*f*) *eight-arm star. After Tagg.*[2]

The use of buried metal strips, rather than wire, does not warrant its extra cost. The equivalent circumference of wire conductors is equally effective.

Buried plates have been used for ground systems, but a plate concentrates more metal at one spot in the earth than is necessary and does not extend its contact over a large volume of earth as do multiple ground rods or buried wires. Accordingly a plate ground is limited in effectiveness and is relatively uneconomical.

For grounding purposes one may wish to connect to a water main providing the pipe is of cast iron or other metal. It could then be regarded as a horizontal conductor buried several feet below the surface. The resistance to earth of the water main and its branches could be calculated for some assumed or measured values of earth resistivity, but it would be preferable to measure it. To estimate what order of resistance one can expect it is

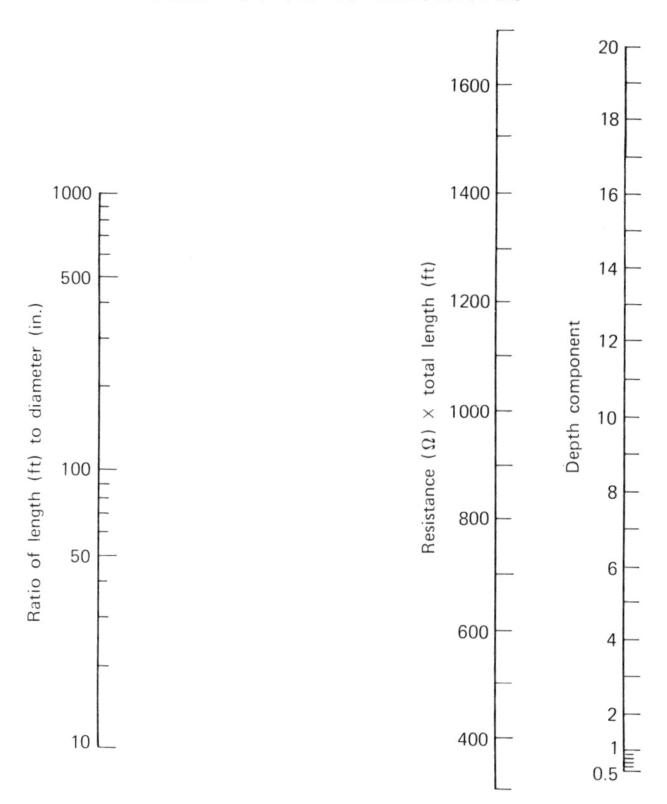

Fig. 4.24. *Nomogram used to obtain resistance of buried wires. After Tagg.*[2]

possible to extrapolate from curves published by Tagg.[2] Let us assume that there are 300 ft of 12-in. cast-iron pipe buried 6 ft below the surface in rocky soil material for which the average resistivity is 2000 Ω-m. From the curve of Fig. 4.22 it will be seen that a 300 ft-conductor that has a 12-in. diameter and is buried 3 ft has a resistance of 1.5 Ω for $\rho = 100$ Ω-m. So for $\rho = 2000$ Ω-m the resistance of the water-main section would be about 30 Ω.

It will be noted that for lengths greater than 300 ft the resistance reduces rather slowly. To take into account the depth of the pipe in our example (i.e., 6 ft) let us refer to Figs. 4.23 and 4.24. From Fig. 4.23 the depth factor is 2. Then using this value in the nomogram of Fig. 4.24, together with a ratio of 25 for the length (feet) to diameter (inches), a value of 480 is obtained for the product of resistance and the length of the pipe. So for the 300-ft length of pipe the resistance component is 1.6 for $\rho = 100$ Ω-m. Converting

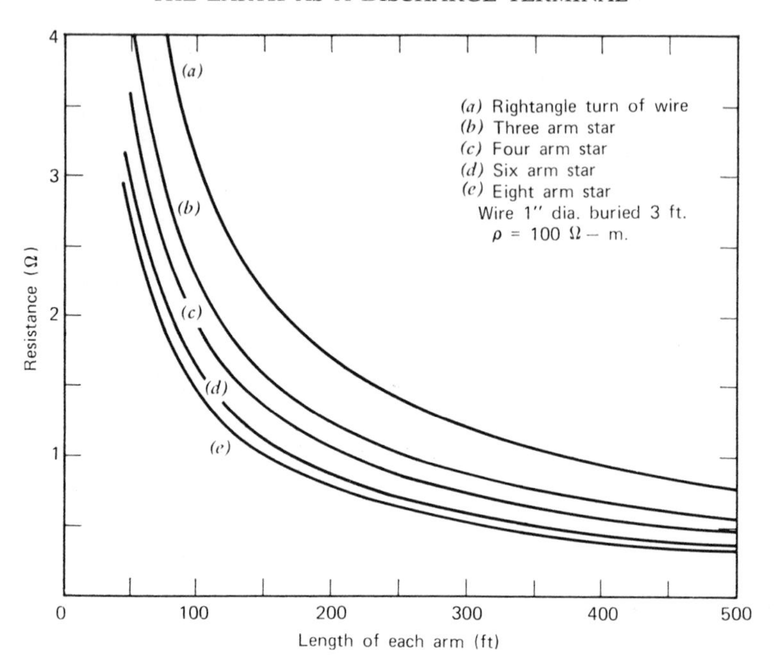

Fig. 4.25. *Resistance (for ρ = 100 Ω-m) of 1-in. diameter wires buried 3 ft deep and arranged as follows: (a) right-angle turn; (b) three-arm star; (c) four-arm star; (d) six-arm star; (e) eight-arm star. After Tagg.*[2]

the resistivity to 2000 Ω-m for the rocky soil material in our example the estimate of resistance becomes 32 Ω. From these two estimates one could use the figure 30 Ω ± 10% as an initial design figure.

4.14 BURIED ELECTRODES IN NONHOMOGENEOUS EARTH

In specifying a grounding system where the resultant resistance must be known to a high degree of certainty, it is first necessary to make earth-resistivity measurements using electrodes at varying depths. In nonhomogeneous soil the earth resistivity will vary with depth, and earth electrodes should be put at depths that are the best compromise between low resistance and economy.

Calculations of the earth resistivity of an electrode buried in an earth layer of one value of resistivity overlaying a layer of different resistivity are complex. Tagg[2] has done this for rods, for horizontal conductors, and also for rods that penetrate into the second layer. The effect of a layer of considerably different resistivity from the surface layer and located well below the electrode has not a large effect on the resistance of the electrode. For example, a 1-in. rod driven 10 ft into ground with a resistivity of 1000 Ω-m extending to a great depth will have a resistance of 300 Ω. If a second layer with a resistivity of 5000 Ω-m is located 20 ft down from the surface, it would increase the effective resistance of the rod by about 30 Ω. On the other hand, if the lower layer had one-fifth the resistivity of the upper layer (i.e., 200 Ω-m), the effective resistance of the rod would be reduced by about 15 Ω.

Now consider a 10-ft rod penetrating two 5-ft layers of earth, where the top layer has a resistivity of 100 Ω-m and the lower layer five times this value. This would add 5Ω to the rod resistance that would be obtained if the surface soil were very deep; that is, the electrode resistance would increase from 30 to 35 Ω for the two-layer case. In the inverse case, where the top 5-ft layer has a resistivity of 500 Ω-m and the lower layer has one of 100 Ω-m, the resistance of the 10-ft rod would be 10 Ω less than the value in deep soil,with a resistivity of 500 Ω-m, that is, 140 instead of 150 Ω. It is apparent that the top layer of soil has the most influence on electrode resistance, and a second layer of equal or lesser depth in contact with the rod has only a minor effect.

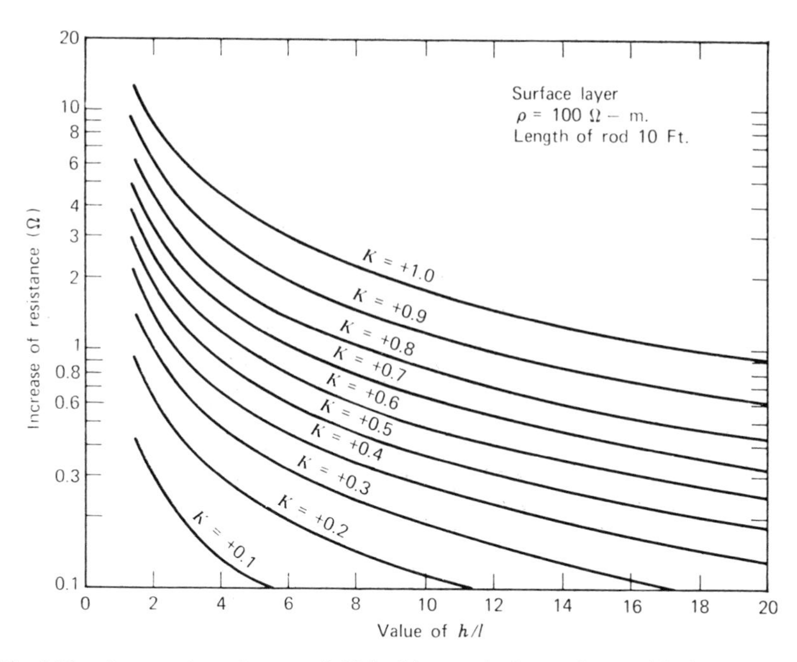

Fig. 4.26. *Increase in resistance of 10-ft driven rods due to layer of higher resistivity (surface-layer resistivity 100 Ω-m). See equation 1 in text for K, the reflection coefficient. After Tagg.*[2]

It is more usual for the lower soil layer to have a higher resistivity than the surface layer. Curves[2] giving the resistance increment of a 10-ft rod penetrating two such layers are shown in Fig. 4.26. The factor K is the reflection coefficient,

$$K = \frac{\rho_2 - \rho_1}{\rho_2 + \rho_1} \tag{1}$$

where ρ_1 and ρ_2 are the resistivities of the upper and lower layers, respectively. The parameter h/l is ratio of the depth of the top layer, h, to the length of the rod, l (10 ft).

Similar curves[2] for buried horizontal wire or strip are shown in Fig. 4.27. In this figure h is the depth of the top layer in which the conductor is buried and l is the horizontal length of the conductor in the same units. In general, test measurements are more reliable than calculations of electrode resistance in soils, but for initial design purposes curves like those calculated by Tagg[2] are available.

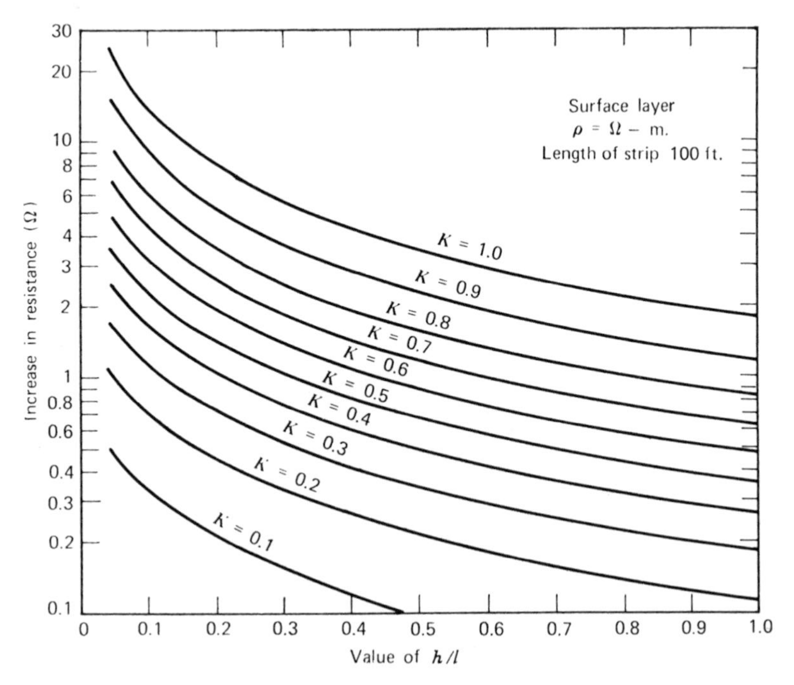

Fig. 4.27. *Increase in resistance of 100-ft buried strip due to lower layer of higher resistivity (surface-layer resistivity 100 Ω-m). See equation 1 in text for K, the reflection coefficient. After Tagg.*[2]

4.15 GROUNDING GRIDS FOR ELECTRIC POWER STATIONS

The objectives of a ground system may be summarized as follows:

1. To dissipate the energy of overvoltages while keeping the potential rise in the earth to a minimum.
2. To prevent damage to electrical apparatus.
3. To safeguard personnel during fault conditions by limiting step and touch potentials.

The potential rise in the earth and overvoltages on equipment are controlled by the ground resistance. The step and touch potentials depend on the voltage gradient, which can be limited by having sufficient horizontal conductors in the buried grid.

For a given area of a grounding system, it has been found that a rod-bed system driven to normal depths of say 10 ft and a wire-grid system buried to a depth of say 3 ft have a combined resistance that is very little different from that of the systems alone. Therefore where the terrain permits a wire grid is adequate and preferable because it equalizes the potential of the area. Of course, where there is a risk of the top layer of soil drying out, a rod bed would ensure a more nearly constant value of system resistance. The resistance of a system of rods of any shape can be simplified to[4]

$$R = \frac{\rho}{2\pi n l_1}\left[\log_e \frac{4l}{a} - 1 + \frac{2k_1 l_1}{\sqrt{A}}(\sqrt{n} - 1)^2\right] \tag{2}$$

where l_1 = length of rod,
$2a$ = diameter of rod,
n = number of rods,
A = area of ground system,
k_1 = a factor depending on the shape of the area and density of the rods.

Also the resistance of a buried wire grid has been simplified to

$$R = \frac{\rho}{\pi l}\left(\log_e \frac{2l}{a^1} + k_1 \frac{l}{\sqrt{A}} - k_2\right) \tag{3}$$

where the symbols have the same meaning and k_1 and k_2 depend on the shape and density of the conductors. The combined resistance of two systems is given by

$$R = \frac{R_{11}R_{22} - R_{12}^2}{R_{11} + R_{22} - 2R_{12}} \tag{4}$$

where R_{11} and R_{22} are the resistance values for the individual systems, and R_{12} is the mutual resistance between them, and

$$R_{12} = R_{21} = \frac{\rho}{\pi l}\left(\log_e \frac{2l}{l_1} + k_1 \frac{l}{\sqrt{A}} - k_2 + 1\right) \tag{5}$$

In general, where the ratio $R_{12}/R_{11} = 0.9$, a second system would only reduce the total resistance by 2% provided $R_{22}/R_{11} > 1.5$ and would reduce it only 5% if $R_{22} = R_{11}$. This indicates that in some cases little would be gained by adding a rod bed to the wire-grid system.

A further simplification of a ground grid system is possible[3,15] by using Laurent's formula

$$R = R_1 + r^1 = \frac{\rho}{4r} + \frac{\rho}{l} \tag{6}$$

where $R_1 = \rho/4r$ = contribution of the grid area to resistance (ohms), $r^1 = \rho/l$ = contribution of the buried conductors to the ground resistance (ohms), r = radius of circle equivalent in area to that of the grid (meters).

From $r = \rho/4R_1$ the area of the grid can be expressed as $A = \pi\rho^2/(4R_1)^2$ m^2. This corresponds to the area given by the ionization radius of Section 4.16 if one makes the length of the rod electrode unity. In Fig. 4.28 the recommended grid areas to give the required values of R_1 are shown by curves for several values of ρ.

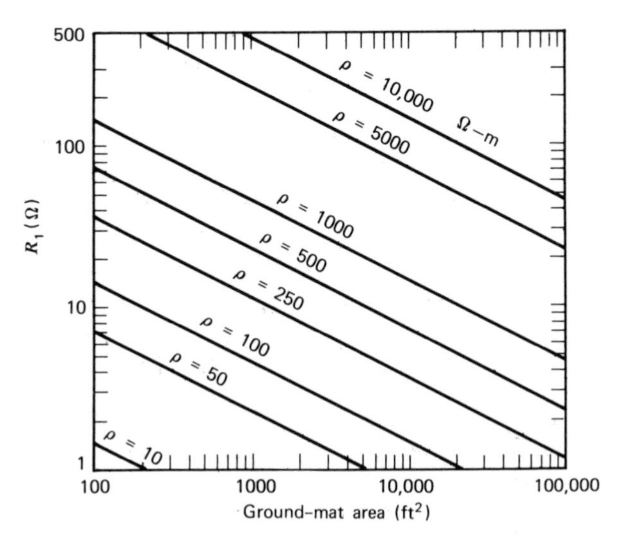

Fig. 4.28. *Recommended grid areas for obtaining required values of R_1 for various values of ρ (Ω-m). After Mukhedkar and Demers.*[3]

The total length of conductor l required to ensure a suitably low voltage gradient for some value of fault current I can be expressed[3] as

$$l = 3.28 \frac{k_1 k_2 \rho I \sqrt{t}}{165 + 0.25\rho_s} \quad \text{feet.}$$

Here k_1 depends on the density of rods in the associated rod bed and their size and depth and k_2 is a correction factor for the nonuniformity of current flow in different parts of the grid. The product $k_1 k_2$ is assumed[3] to be 1.25. The quantity ρ_s is the resistivity of the surface material (e.g., crushed stone) and may be assumed to be 4000 Ω-m, and t is the duration of the discharge-current surge (assumed here to be 0.5 sec). Substituting these values, the length l becomes $2.49 \times 10^{-3}\ \rho I$ feet. This relationship is shown in Fig. 4.29.

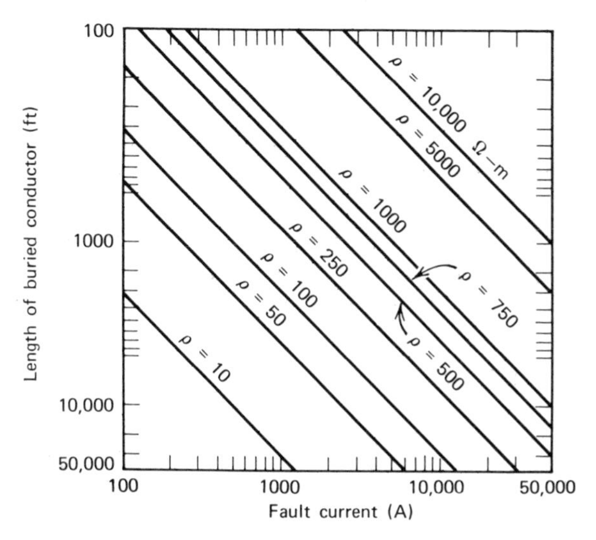

Fig. 4.29. *Relationship between length of buried conductor and fault current for various values of ρ (Ω-m). After Mukhedkar and Demers.*[3]

The quantity $r^1 = \rho/l$ can now be expressed as $r^1 = (1.32 \times 10^3)/I$, and it is usually relatively small.

To work out an example, let us assume the following conditions:

$\rho = 500$ Ω-m.
Desired value of R: 5 Ω.
Fault current $I = 5000$ A.
The value of r^1 is $(1.32 \times 10^3)/5000 = 0.26$ Ω.

So the required value of R_1 is approximately 4.7 Ω. The recommended grid area to provide an R_1 value of 4.7 Ω for $\rho = 500$ Ω-m can be read from Fig. 4.28 as 20,000 ft². The total length of conductor, l, to provide a suitably low voltage gradient is, from Fig. 4.29, 4000 ft.

Before considering the associated rod bed, let us estimate the resistance of the grid itself, assuming that it is buried about 3 ft deep. From the formula of Schwartz[6]

$$
\begin{aligned}
R &= \frac{\rho}{l}\left(\log_e 6l - 5.6 + 1.4\frac{l}{\sqrt{A}}\right) \\
&= \frac{500}{4000}\left(\log_e 24{,}000 - 5.6 + 1.4\frac{4000}{142}\right) \\
&\simeq 5.5\ \Omega.
\end{aligned}
$$

That is, the resistance of the grid alone is nearly low enough. However, it has been found[4] that the resistance of combined systems of grids and rod beds is only slightly less than that of one of the systems alone. Therefore, in our example, we would need a rod bed whose resistance is about 4.7 Ω to achieve the objective exactly. From Fig. 4.20 it is apparent that 50 rods (10 ft long, 0.75-in. diameter) would be required. Let us say 48 rods, in six rows of 8 rods each. The grid mat would be like that sketched in Fig. 4.30, but other shapes could be used. The rods can be spaced 23 ft apart, which is

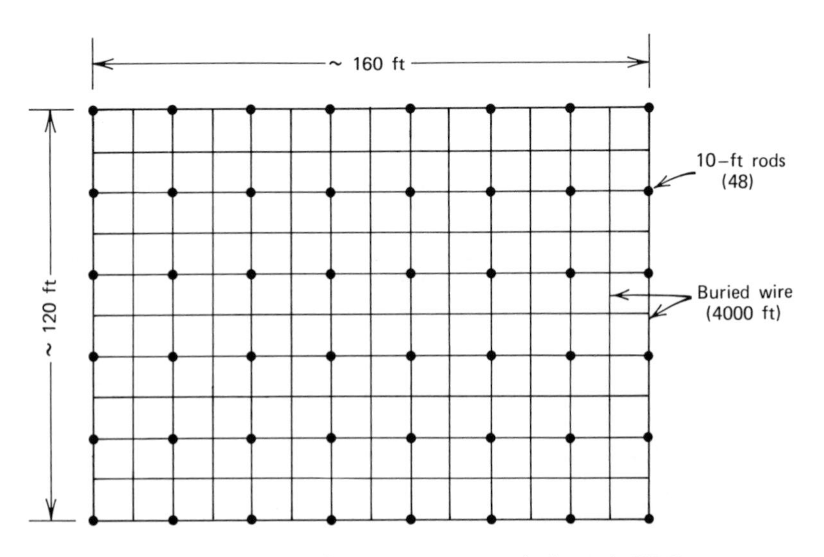

Fig. 4.30. *Ground grid of 5-Ω resistance, in soil where ρ is 500 Ω-m.*

sufficient for 10-ft rods. In order to have about 4000 ft of horizontal conductor it is necessary to have additional buried wire between each row of rods, so that the mesh of the grid is 11 ft (approximately) square. It is interesting to note that this size of grid provides a resistance that is one-hundredth the value of the earth resistivity in ohm-meters. Accordingly, if this ground grid were in soil where ρ is 1000 Ω-m, for example, its resistance would be 10 Ω.

4.16 SOIL IONIZATION

If a sufficiently high lightning voltage reaches an earth electrode, such as a ground rod, the soil will ionize out to some distance from the conductor. The ionization-voltage gradients for several soil types are shown in Table 4.4. These should be regarded as approximate values. Most lightning-discharge voltages exceed those shown in the table, so that for single earth

TABLE 4.4 CRITICAL DISRUPTIVE GRADIENT OF SOILS[a]

Soil	Gradient (kV/cm)
Gravel, moist:	
To produce breakdown on tail of wave	11.4–12.3
To produce breakdown on crest of wave	17.8–19.2
Gravel, dry:	
Breakdown on crest of wave or not at all	20.8–22.8
Sand, moist:	
To produce breakdown on tail of wave	13.0–16
To produce breakdown on crest of wave	17.1–23.4
Sand, dry:	
Breakdown on crest of wave or not at all	17.1–18.8
Clay, plastic:	
Time to breakdown erratic from 1 to 8 μsec[b]	18.7–39.0

[a] 1.5 × 40 wave.
[b] Wave departed from standard.

electrodes ionization of the soil surface would occur. Its effect is to lower the voltage at the electrode to the value required for ionization. Fortunately the ionization-voltage gradient in sand and gravel is relatively low, which compensates somewhat for their high resistivity.

If the soil is ionized by the strike voltage out to a radius A_0 from the rod, the total current flowing from the rod is spread over the surface of a cylinder of that radius so as to produce a current density of $I/2\pi A_0 L$ amperes per square meter, where L is the depth of the rod. The potential gradient at the cylinder's surface is then $E_0 = I\rho/2\pi A_0 L$, and this is the breakdown-potential gradient. Alternatively the ionizing radius can be expressed as a function of the breakdown potential: $A_0 = I\rho/2\pi L E_0$. The impedance of the rod to ground under these conditions is called the impulse impedance and is expressed by

$$R_0 = \frac{\rho}{2\pi L}\left(\log_e \frac{4l}{A_0} - 1\right)$$

This is a smaller value than that for the nonionized case,

$$R = \frac{\rho}{2\pi L}\left(\log_e \frac{4l}{a} - 1\right)$$

because it applies to the larger surface at radius A_0. A comparison between these two values, for specific rod dimensions and earth resistivity, is shown in Fig. 4.31. As the crest current of the stroke passes the fall in ionization lags behind the current decay, but in time the effective radius shrinks to the real radius of the rod. It follows that the larger the radius of a ground system,

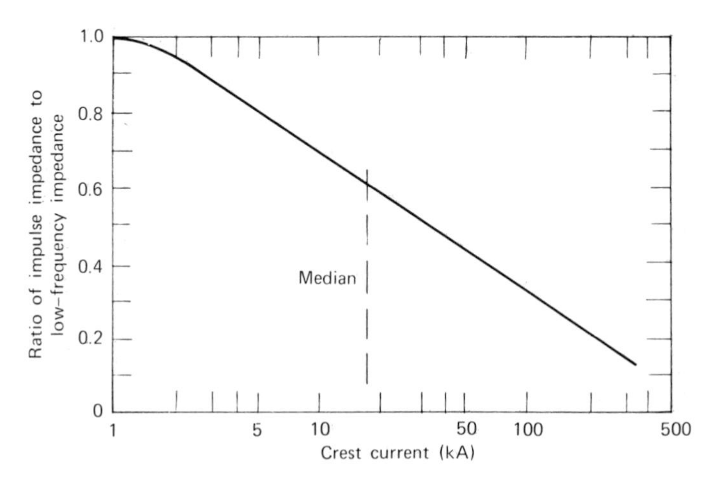

Fig. 4.31. *Ratio of impulse (surge) impedance to Low-frequency impedance for 2-m ground rod (ρ = 100 Ω-m, a = 0.625 cm).*

the smaller will be the difference between the ionized value of impedance and the nonionized value.

The curve in Fig. 4.32 shows the ratio of impulse impedance (largely resistive) to the low-frequency impedance for a rod in good soil. The effect of various soil types on impulse resistance is illustrated in Fig. 5.8.

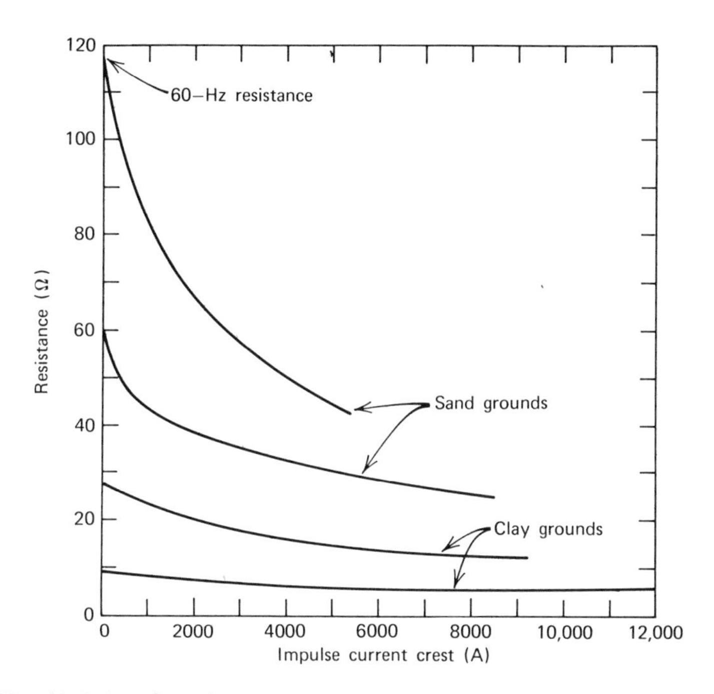

Fig. 4.32. *Variation of impulse resistance with impulse current for various values of 60-Hz resistance. After Ref. 6.*

APPENDIX 4.A

DERIVATION OF FALL-OF-POTENTIAL METHOD

For an analysis of the fall-of-potential method, let the electrode E be a hemisphere of radius r and the other two electrodes designated as shown in Fig. 4.6. The current passing between electrodes C and E, which are separated by a distance D, is I. The potential at hemisphere E due to entering current can be expressed as $I\rho/2\pi r$, and the potential at C due to leaving current is $V = I\rho/2\pi(D - r)$. This results in a total potential at E of

$$V = \frac{I\rho}{2\pi}\left(\frac{1}{r} - \frac{1}{D - r}\right)$$

Similarly the potential at E with respect to the intermediate electrode P at distance H is

$$V = \frac{I\rho}{2\pi}\left(\frac{1}{H} - \frac{1}{D - H}\right)$$

From this the net potential difference between E and C becomes

$$V = \frac{I\rho}{2\pi}\left(\frac{1}{r} - \frac{1}{D - r} - \frac{1}{H} + \frac{1}{D - H}\right) \tag{A.1}$$

and the measured resistance of electrode E is $R = V/I$, that is, expression A.1 divided by I.

Further consideration of expression A.1 can lead to a unique value for the positioning of the potential electrode P for an accurate measurement of electrode resistance by the fall-of-potential method. As the radius r is very small compared to distance D, it can be dropped from the $1/D - r$ term of formula A.1. Then

$$R = \frac{\rho}{2\pi}\left(\frac{1}{r} - \frac{1}{D} - \frac{1}{H} + \frac{1}{D - H}\right) \tag{A.2}$$

If the ratio D/r is called c and H/r is called p,

$$R = \frac{\rho}{2\pi r}\left(1 - \frac{1}{c} - \frac{1}{p} + \frac{1}{c - p}\right) \tag{A.3}$$

Previously it was found that $R_\infty = \rho/2\pi r$; hence

$$\frac{R}{R_\infty} = \left(1 - \frac{1}{c} - \frac{1}{p} + \frac{1}{c - p}\right) \tag{A.4}$$

The aim now is to make the measured R equal to R_∞, and this would occur for

$$\left(-\frac{1}{c} - \frac{1}{p} + \frac{1}{c - p}\right) = 0$$

This can be rearranged in the form

$$p^2 + cp - c^2 = 0 \tag{A.5}$$

$$p = \frac{-c \pm \sqrt{5c^2}}{2} \tag{A.6}$$

and for p positive,

$$p = c\left(\frac{\sqrt{5} - 1}{2}\right) = 0.618c \tag{A.7}$$

From this it follows that $H = 0.618D$. This indicates that for any separation of the current electrodes the true resistance of one of them is obtainable when the potential electrode is at 61.8% of the distance toward the other. As this derivation assumes a hemispherical electrode, it is necessary to use the equivalent hemisphere of any electrode or system of electrodes.

APPENDIX 4.B

SINGLE ROD—RESISTANCE TO EARTH AND EQUIVALENT RADIUS

In Fig. 4.8 is sketched a cylindrical rod of length l and diameter $2a$, with its image above the surface of the earth.

A uniform charge over its surface is assumed, and its capacity C is to be calculated. If the charge is q per centimeter of length, the charge on the band dy is $q\,dy$, and the resulting potential at any point P can be expressed, to a close approximation, as

$$\frac{q\,dy}{a^2 + y^2}$$

To obtain the total potential this expression must be integrated over the range from $y = 0$ to $y = l + x$. The result is then multiplied by dx/l and integrated from $x = 0$ to $x = l$. This gives the average potential:

$$\frac{V}{2q} = \log_e \frac{4l}{a} - 1 \tag{B.1}$$

From this the capacity C can be calculated. The reciprocal of capacity is

$$\frac{1}{C} = \frac{V}{2ql} = \frac{1}{l}\left(\log_e \frac{4l}{a} - 1\right) \tag{B.2}$$

The resistance is

$$R = \frac{\rho}{2\pi C} = \frac{\rho}{2\pi l}\left[\log_e \frac{4l}{a} - 1\right] \tag{B.3}$$

If the diameter $2a$ is called d, then

$$R = \frac{\rho}{2\pi l}\left(\log_e \frac{8l}{d} - 1\right) \tag{B.4}$$

The resistance of a hemispherical electrode of radius r is $\rho/2\pi r$. Equating this to the resistance of the rod,

$$\frac{\rho}{2\pi r} = \frac{\rho}{2\pi l}\left(\log_e \frac{8l}{d} - 1\right) \tag{B.5}$$

Then the equivalent radius is obtained:

$$r = \frac{l}{\left(\log_e \dfrac{8l}{d} - 1\right)} \tag{B.6}$$

APPENDIX 4.C

THREE RODS IN PARALLEL IN A STRAIGHT LINE

Consider three buried rods in a line, separated by distance d, as shown in Fig. 4.11. Being connected in parallel, they will be at the same potential, but the charge on the center rod will not be the same as that on the two outer rods, which have some influence on it. Let the charge on the outer rods be Q_1 and that on the center rod Q_2. The potential on each outer rod is given by

$$V_1 = \frac{Q_1}{r} + \frac{Q_2}{d} + \frac{Q_1}{2d} = \frac{1}{r}\left[Q_1\left(1 + \frac{\alpha}{2}\right) + Q_2\alpha\right] \tag{C.1}$$

and the potential on the center rod is

$$V_2 = \frac{Q_2}{r} + \frac{2Q_1}{d} = \frac{1}{r}\left(Q_1 2\alpha + Q_2\right) \tag{C.2}$$

As V_1 and V_2 are the same potential, these two expressions are equal:

$$Q_1\left(1 + \frac{\alpha}{2}\right) + Q_2\alpha = 2\alpha Q_1 + Q_2 \tag{C.3}$$

$$Q_2 = Q_1 \frac{(1 - 3\alpha/2)}{(1 - \alpha)} = KQ_1 \tag{C.4}$$

where K represents the bracketed terms.

The potential can now be expressed as

$$\frac{1}{r}(Q_1 2\alpha + KQ_1) = \frac{Q_1}{r}(2\alpha + K) \tag{C.5}$$

The total charge is $2Q_1 + Q_2$ or $Q_1(2 + K)$. Then to find the capacity C one can write

$$\frac{1}{C} = \frac{\text{potential}}{\text{total charge}} = \frac{1}{r}\frac{2\alpha + K}{2 + K} \tag{C.6}$$

and from

$$R_{\text{total}} = \frac{\rho}{2\pi C} \tag{C.7}$$

$$R = \frac{\rho}{2\pi r}\frac{2\alpha + K}{2 + K} \tag{C.8}$$

Remembering that the resistance of one rod is $\rho/2\pi r$, the ratio of the combined resistance to that of one rod is $(2\alpha + K)/(2 + K)$.

APPENDIX 4.D

RESISTANCE OF BURIED HORIZONTAL WIRES

Assume a buried wire of length $2l$, buried a depth of $S/2$, and an image of its charge at $S/2$ above the earth's surface, as indicated in Fig. 4.33. The potential of the wire due to its own charge q is[2]

$$V_1 = 2q\left(\log_e \frac{4l}{a} - 1\right) \tag{D.1}$$

where a is the radius of the wire.

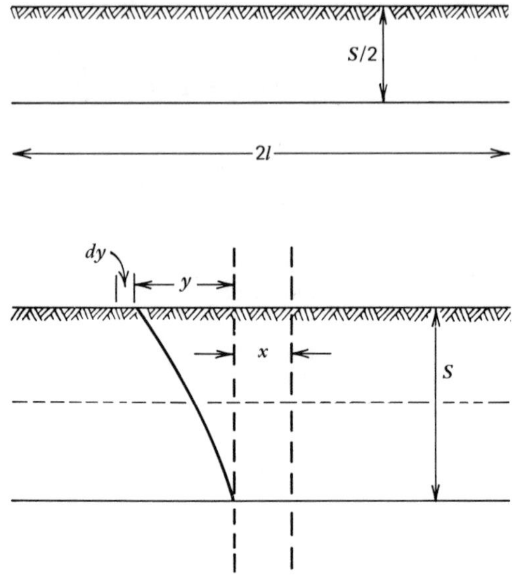

Fig. 4.33. *Buried horizontal wire—derivation of resistance.*

To find that portion of the potential on the wire due to its image charge, one may start with the potential at any point on the buried wire at some distance x from its center produced by an element of image charge $q\,dy$. This element of potential is $q\,dy/(s^2 + y^2)$.

The potential at the point due to the whole image is the integral of this expression over the length of the wire. The result is then added to the potential due to the wire charge itself to give the total potential:

$$2q\left[\log_e \frac{4l}{a} - 1 + \log_e \frac{2l + \sqrt{S^2 + 4l^2}}{S} + \frac{S}{2l} - \frac{\sqrt{S^2 + 4l^2}}{2l}\right] \quad \text{(D.2)}$$

If the total potential is now divided by the total charge $4ql$ on the wire and image, the result is $1/C$, where C is the capacity. Then using the relation $R = \rho/4\pi C$, the expression for resistance to earth is

$$R = \frac{\rho}{4\pi l}\left[\log_e \frac{4l}{a} - 1 \right.$$

$$\left. + \log_e \frac{2l + \sqrt{S^2 + 4l^2}}{S} + \frac{S}{2l} - \frac{\sqrt{S^2 + 4l^2}}{2l}\right] \quad \text{(D.3)}$$

REFERENCES

1. E. Hanle, "The Complex Impedance of the Earth's Surface at Radio Frequencies and Its Measurement," *NTZ-Commun. Journal,* No. 3 1966.
2. C. F. Tagg, *Earth Resistance,* Newnes, London, 1964.
3. D. Mukhedkar and G. Demers, "Ground Grid Design," *Eng. J. EIC,* **51/2,** February 1968.
4. S. J. Schwartz, "Analytical Expressions for the Resistance of Grounding Systems," *Power Apparatus and Systems,* No. 13, August 1954.
5. AIEE Working Group No. 56.1, *Guide for Safety in AC Substation Grounding,* AIEE Publ. No. 80.
6. Engineers of Westinghouse Electric Corporation, *Electrical Transmission and Distribution Reference Book,* 4th ed., 1964, Chapter 17.

BIBLIOGRAPHY

C. L. Roach, "Microwave Tower Grounding Systems," paper presented at C.E.A. Meeting, Montreal, January 1964.

E. D. Sunde, *Earth Conduction Effects in Transmission Systems,* D. Van Nostrand, Princeton, N.J., 1949.

CHAPTER 5

Protective Grounding Systems—General

5.1 OBJECTIVES

The objectives in lightning protection are to save property from damage or loss by fire; to prevent the disruption of essential services, such as electrical power and telephone communications; and to protect human and animal life.

The need for protection at a particular structure or building can be determined by considering objective, subjective, and economic reasons. Objective factors are the disruption of a business or activity; the loss of essential services, such as electricity or telephone; and the danger to employees. Subjective reasons are public safety, the historic value of a structure, and reassurance of owners or tenants. The economic factor is the cost of insurance to cover losses compared with the cost of providing lightning protection.

5.2 MEASUREMENTS FOR THE DESIGN OF GROUNDING SYSTEMS

The principles of earth-resistivity measurements were described in Section 4.3 and the principle of measuring the resistance to earth of a buried electrode or of a ground system by the fall-of-potential method was described in Section 4.5. Information on the practical aspects of earth measurements is available[1,2] from several sources, including instruction booklets with the instruments themselves. A few general words of caution might, however, be appropriate. In built-up areas there might exist stray currents from other electrical systems. Such currents can produce erroneous readings, but with a suitable measuring instrument and correct methods the effect can be minimized. The existence of buried pipes or other metal can distort soil-resistivity measurements. By taking sufficient measurements at large spacings and in directions at right angles to the initial tests such discrepancies can be revealed. Buried metal objects can influence the resistance value of a buried ground system, but this is of no consequence provided the resistance of the ground system is measured at its terminals where the lightning-discharge current would enter.

The design of grounding systems can be facilitated by earth-resistivity measurements and by taking note of the measured results of installed systems. The earth resistivity varies with moisture content and temperature, but in many cases, if the measurement probes are spaced about 100 ft apart, the resistivity value obtained will be relatively independent of the seasonal variations. Furthermore, this resistivity value can be used for design purposes with considerable reliability.

However, if there is bedrock beneath the layer of soil, the resistivity will show a minimum value for a probe spacing equal to the soil depth and will then increase with a greater spacing of the probes. In this case any grounding system in the soil will have a resistance compatible with the minimum measured value of resistivity. In measuring earth resistivity with a wide probe spacing the resistance reading on the measuring instrument might be a low value, leading to inaccuracy. This can be overcome by spacing the potential probes wider apart, as long as they stay equidistant from the center of the line of probes. With the arrangement of probes illustrated in Fig. 5.1, the expression for resistivity is

$$\rho = \pi R \frac{(A^2 + AD)}{D}$$

where D is the spacing between the potential probes and A is the spacing between each potential probe and the corresponding current probe.

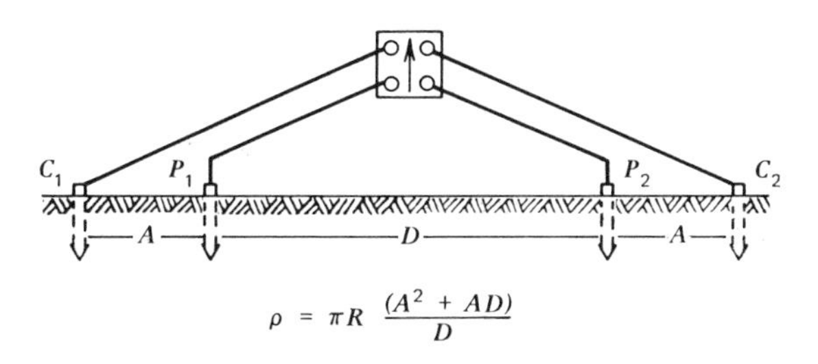

Fig. 5.1. *Method for increasing potential reading.*

Test measurements should be taken along the lines of a grid, starting with probe spacings of 10 ft and increasing in steps of 10 ft to a spacing of 100 ft, over the area of the intended ground system. For example, measurements may be made along three equispaced parallel lines in one direction, with a similar set of measurements being made in a perpendicular direction.

In earth of low resistivity that appears to be uniform the potential probes may be placed at right angles to the current probes.[2] This permits a closer spacing of the probes, for a suitably high resistance reading on the instrument.

A valuable comparison between design values and measured values of ground-system resistance for five power substations[1] is reproduced in Table 5.1.

TABLE 5.1 COMPARISON OF COMPUTED AND MEASURED SUBSTATION GROUND-GRID RESISTANCE[a]

	Substation				
Parameter	A	B	C	D	E
Resistivity proposed for design (Ω-cm)	200,000	80,000	20,000	130,000	2800
Area of grid (ft²)	15,159	60,939	18,849	15,759	61,479
Equivalent radius (ft)	69.5	139.2	77.5	71.0	140.0
Buried conductor (ft)	3120	9500	1775	3820	3000
Calculated resistance (Ω)	25.7	4.97	2.55	16.15	0.191
Measured resistance (Ω)	39.0	4.1	3.65	18.2	0.21
Soil texture	Sand and gravel	Sandy loam	Sand and clay	Sand and gravel	Soil and clay

[a] Data from Kinyon.[1]

These ground systems were horizontal conductors buried 12 in., intersperced with 8- and 10-ft ground rods. There is remarkably good agreement between computed values of ground-system resistance and the measured values, except in one case where the soil was sand and gravel. It is worth noting, too, that in sand and gravel it is not possible within practical limits to lower the ground resistance to a few ohms.

5.3 INSTRUMENTS FOR MEASURING EARTH RESISTANCE

Detailed information on measuring instruments is available from several sources[2,3] and from manufacturers. Such information will not be repeated here, but a brief description of the main types will be given. The use of the instruments is not difficult, and with proper precautions reliable measurements can be obtained.

OHMMETER TYPE

This is the usual kind of instrument used in earth measurements. The voltage is generated by a hand-driven generator, battery-operated vibrator, or transistor converter. A voltage in the 500-V range is common, and the current is alternating or reversing direct current, at a rate slightly above the normal power-line frequency. In this way the effects of stray currents in the earth, either direct or at power-line frèquency, are minimized. The application of the instrument is described in Section 4.3. The circuit of a typical instrument is shown in Fig. 5.2.

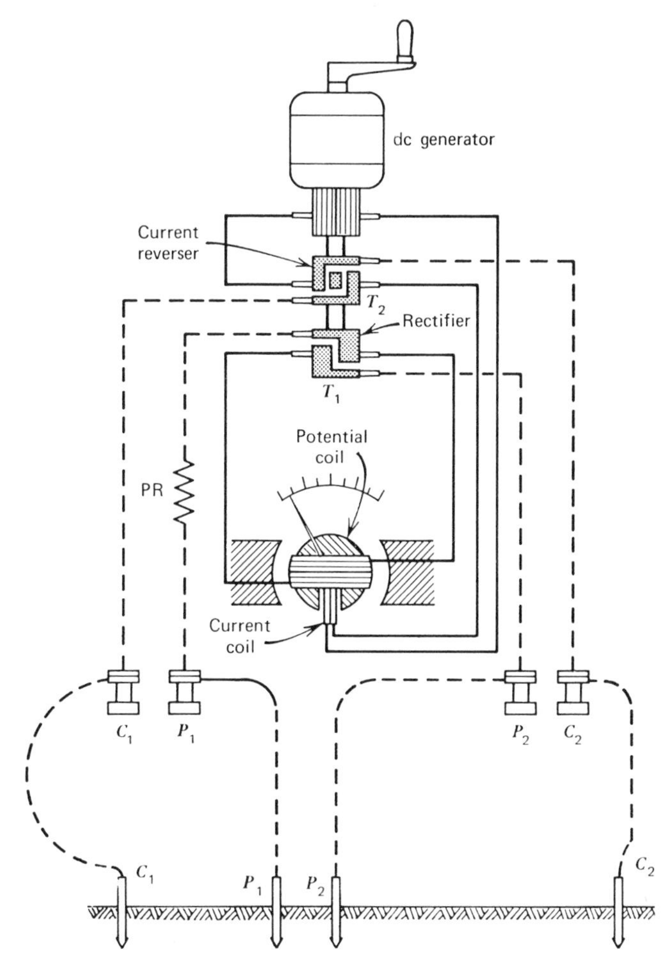

Fig. 5.2. *Ohmmeter type of resistance-measuring instrument. Circuits carrying direct current are indicated by solid lines; those carrying alternating current, by broken lines.*

RESISTANCE COMPARISON OR BRIDGE TYPE

This type of instrument uses a similar source of voltage and oscillatory current as the ohmmeter type of instrument. The circuit is illustrated in Fig. 5.3.

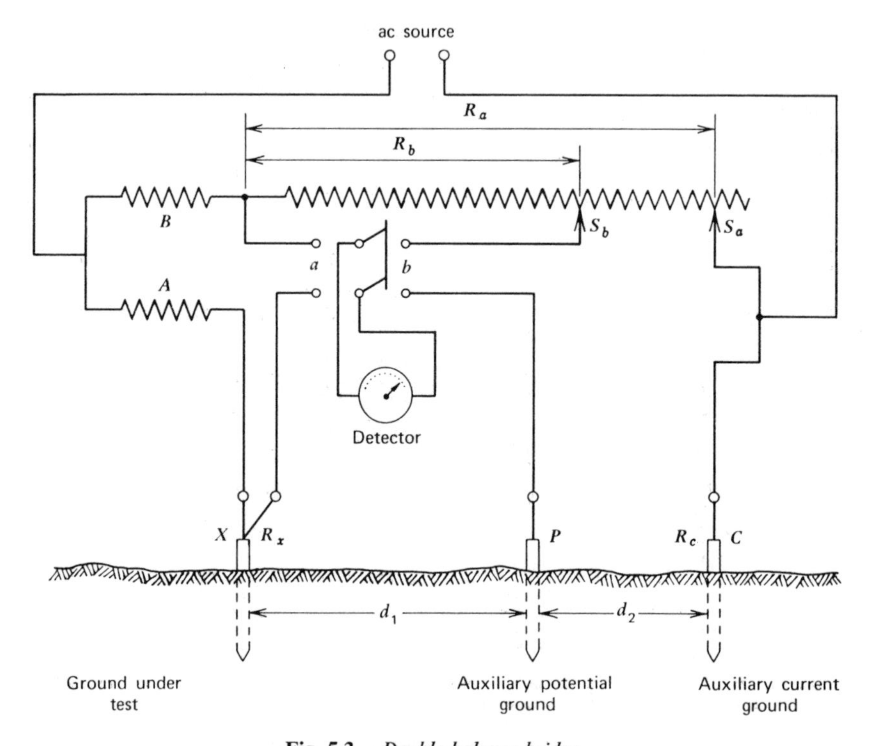

Fig. 5.3. *Double-balance bridge.*

In this method current from the ac source flows in two parallel circuits. The lower circuit includes fixed resistance A, ground X under test, and auxiliary current ground C. The upper circuit includes fixed resistance B and an adjustable slide rheostat on which two sliders S_a and S_b, make contact. With the detector switch closed to the left, slider S_a is adjusted until the detector shows a balance. The currents in the two branch circuits are then inversely proportional to resistances A and B. The switch then is closed to the right, and slider S_b is adjusted until the detector again shows a balance. The potential drop between X and P is then equal to the drop in portion Rb of the slide rheostat, and the resistance of the ground under test then is given by

$R_x = R_b A/B$. The scale over which S_b moves can be calibrated to read R_x directly.

NULL-BALANCE INSTRUMENT

This instrument may have a generator or vibrator current source. A circuit diagram is shown in Fig. 5.4. When the rheostat R_1 is adjusted to produce zero potential in the galvanometer and potential probe circuit, its tapped portion is related to the resistance of the ground under test. The rheostat can be calibrated in ohms of resistance to ground; a range switch is provided by transformer taps in the potential probe arm.

COMBINED OHMMETER AND NULL-BALANCE INSTRUMENT

A more sophisticated instrument combining these two principles is available. Its chief advantage is greater precision in measuring small values of resistance.

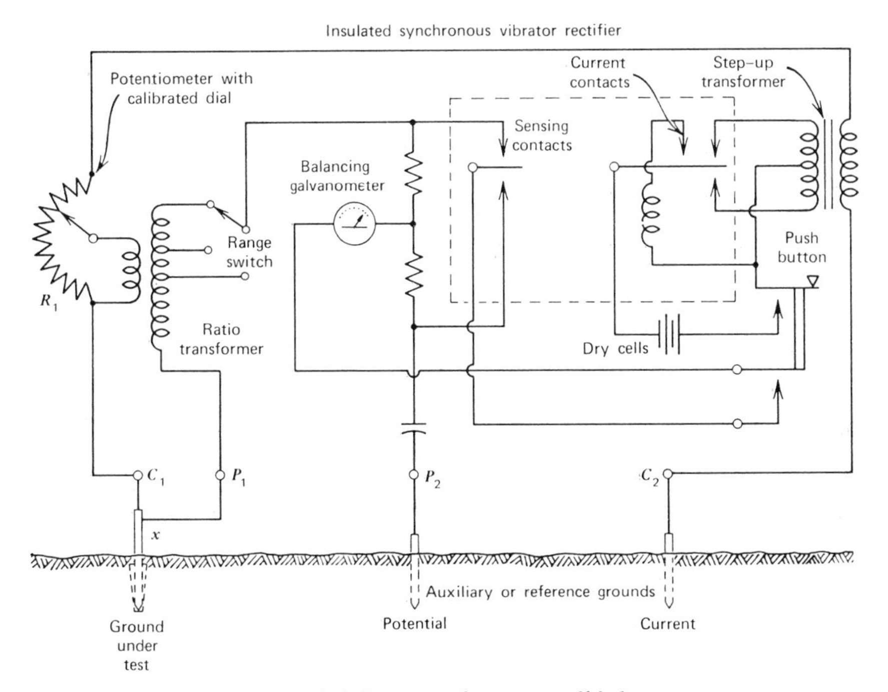

Fig. 5.4. *Single-balance transformer, or null-balance type.*

It is intended primarily for geophysical exploration, but can also be used in measurements for grounding systems.

5.4 PROTECTION PRINCIPLES ABOVE GROUND

Lightning usually terminates on some vertical projection of a structure—except in open country, where it may strike anywhere, but with a preference for high ground or hills if they are in the vicinity of the thundercloud.

In lightning protection it is important to recognize what part of a structure is likely to receive a strike from a thundercloud overhead or to provide a metal rod and/or conducting wire as an "air terminal" for lightning. If the tallest object is a metal mast on, or adjacent to, a structure, this will act as the air terminal, and it should be connected to the grounding system by a down conductor.

In lightning protection one aims to protect only objects that are sufficiently isolated from higher objects to be vulnerable to a strike.

The lowest value of resistance attained in a system of earth conductors depends on the type of terrain, the area of terrain available, and economic limits. A value of ground-system resistance considered acceptable by some electric power systems and protection codes is 10 Ω. At locations in which there is a deep layer of loam it is possible to obtain a lower value than this, whereas at locations in which rock lies close to the surface it would be economically unreasonable to lower the ground-system resistance to 10 Ω. Protection is not absolute, it is a matter of degree. The value of 10 Ω is a recommended nominal objective.

5.5 ATTRACTIVE ZONE

The horizontal distance between the tip of a downcoming leader stroke and a lightning conductor, or air terminal, that receives the strike is called the attractive range, and the area within this range may be called the attractive zone. The attractive range increases as the intensity of the lightning discharge increases, because the striking distance increases. The striking distance of a lightning stroke is the distance from the grounded object upward to the point where the earth streamer from that object meets the leader tip to initiate the stroke proper.

The striking distance varies with the intensity of the stroke. The breakdown-voltage gradient of the air ahead of the tip of a negative downward stepped leader is about 5 kV/cm, and the magnitude of the electric gradient at the leader tip is proportional to the charge on the leader, which in turn determines the stroke-current amplitude. For the average lightning-stroke current of, say, 20 kA the striking distance is about 30 m; for a lightning-stroke current of 40 kA the striking distance would be about 40 m; and for a 100-kA current, about 100 m (see Fig. 5.5).

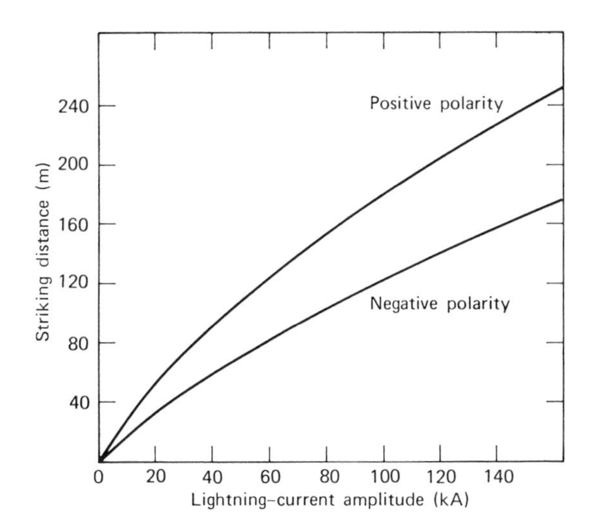

Fig. 5.5. *Striking distances of negative and positive lightning strokes. After Golde.*[4]

In photographs of strokes the meeting point of the downcoming leader tip and upward streamer, sometimes called the strike point, is occasionally visible as a bend in the stroke path where the leader tip diverts toward the streamer or as a downward fork of the luminous path. This is estimated to be at about 60% of the distance from the earth object to where the voltage breakdown occurred at the leader tip. From such observations it is possible to roughly estimate the magnitude of the stroke current.

5.6 PROTECTIVE ZONE

The vulnerability of an object to being struck is largely independent of its insulating properties. For example, the protection afforded to an electric power line by an overhead ground wire is due to its being closer to the thundercloud than the phase wires. A person standing on dry sand in the middle of a desert (assuming a thundercloud develops there) is just as vulnerable as a person standing on a wet raft on the ocean. However, once lightning strikes the nearest object and that object has a high insulating value that prevents the immediate lowering of its potential, a side flash can occur to an adjacent object.

An elevated earth terminal will receive a lightning strike in preference to lower objects around it. That is to say, it provides a protective zone for other objects in the vicinity. This protective zone is not a constant because it depends on the position of the thundercloud and the intensity of the stroke. Protective-zone radii based on experience have been established, but countries differ somewhat in their views as to what this should be. Table 5.2 gives several values for the ratio of protected radius to height of "air terminal," R/H, and in some cases the angle of protection, θ, from the terminal to a protected point on the ground.[5] Two of the examples include limits for particularly dangerous buildings, such as those for storing explosives or flammable material.

TABLE 5.2 PROTECTIVE RANGE AND ANGLE[a]

Country	R/H[b]	θ (degrees)	R/H[b]	θ (degrees)
			Dangerous Structures	
United States	2:1		1:1	45
Britain	1:1	45	0.58:1	30
South Africa	1:1	45		

[a] From Wood.[5]
[b] Ratio of protected radius R to height of air terminal, H.

5.7 EFFECT OF CURRENT WAVE SHAPE ON ELECTRODE IMPEDANCE

The current wave shape of the predominant type of lightning stroke, one that lowers negative charge from the cloud, has a steep front and a relatively long decay interval. A typical waveform, illustrated in Fig. 5.6, rises to its crest value in 1.5 μsec and decays to half the crest value in 40 μsec. It would be described as a 1.5×40 wave.

The impedance of ground rods to such a wave would not differ appreciably from their low-frequency impedance. However, a long buried horizontal wire would have sufficient inductance to raise its impedance significantly. It has been shown[7] that a single buried wire 50 m long would present double the value of its 60-Hz impedance to a 1.5×40 current wave. However, if the ground connection were made to the center of a 100-m buried wire, so

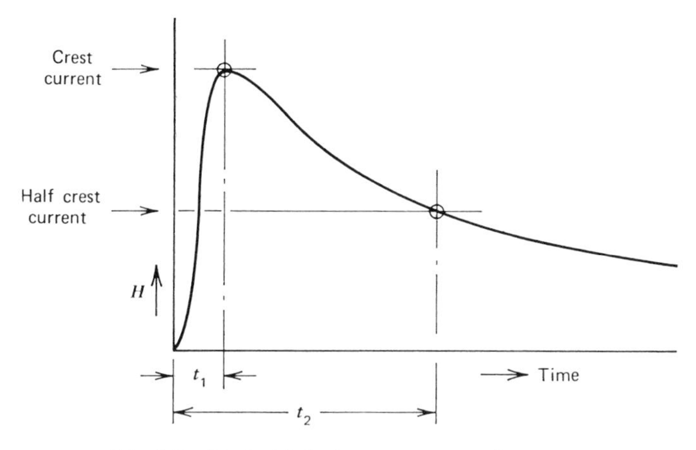

Fig. 5.6. *Typical lightning-current wave shape.*

that there were two 50-m lengths in parallel, the impedance would be reduced to nearly its low-frequency value. This illustrates the advantage of having several short buried radials in a star pattern rather than a single long wire. A group of buried radials also comprises a capacitive admittance which contributes to the reduction of surge impedance. The relationship between the surge impedance and the number of buried radials is shown in Fig. 5.7.

For a given ground-system impedance the surge voltage at its terminal is proportional to the lightning-stroke current. If this voltage surge becomes

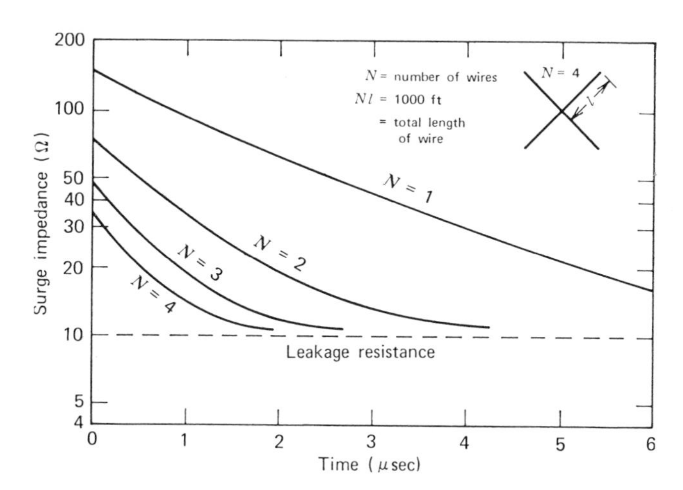

Fig. 5.7. *Surge impedance versus number of buried wires having the same total length. After Bewley et al.*[6]

high enough with respect to the remote earth, the soil in the immediate vicinity will be ionized by the high voltage gradient. The effect of soil ionization is to reduce its resistance to a low value out to a radius where the gradient has decreased below the ionizing or breakdown gradient. If this radius of ionized soil is A_0 meters, surrounding a ground rod of length l meters, there will be an ionized cylinder of earth of $2\pi A_0 l$ square meters. If the total discharge current is I, the current density over this surface will be $I/2\pi A_0 l$, and if the average soil resistivity is ρ ohm-meters, the voltage gradient at the surface of the cylinder will be $E_0 = \rho I/2\pi A_0 l$ volts per meter. This is called the breakdown, or disruptive, voltage gradient for that particular soil. Conversely, if one knows the breakdown-voltage gradient, the ionizing radius for any value of lightning-discharge current can be calculated from $A_0 = I\rho/2\pi l E_0$. Common values of breakdown-voltage gradient for several soil types are given in Table 4.4.

The ionized cylinder of earth acts like a large cylindrical electrode having a lower impedance whose value can be found from the formula for a cylindrical rod:

$$R_0 = \frac{\rho}{2\pi l}\left(\log_e \frac{4l}{A_0} - 1\right) \tag{1}$$

This is called the impulse impedance, as it only applies until the crest voltage falls below the ionizing value. The ratio of impulse impedance to the low-frequency impedance for various soils and ground-rod dimensions is shown in Fig. 5.8 for increasing values of crest current. This illustrates the value of

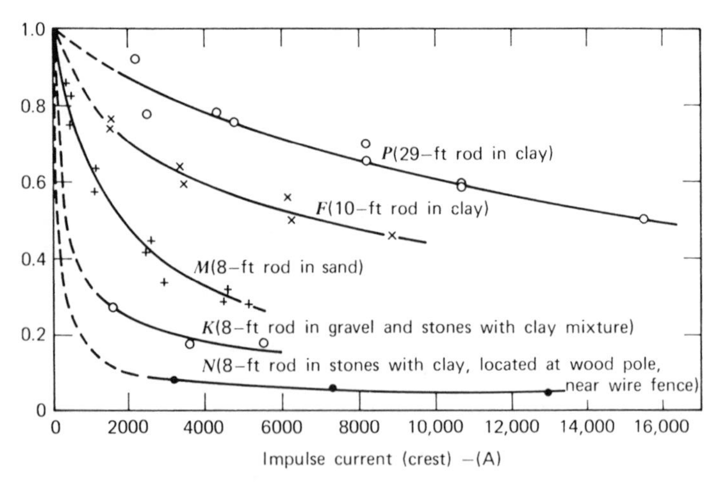

Fig. 5.8. *Ratio of impulse to 60-cycle resistance for various soils and vertical ground rod dimensions as a function of crest current values. After Bellaschi.*[11]

having a vertical depth electrode at the center of a grounding system of horizontal conductors. The addition of buried horizontal ground conductors would have a reducing effect on the voltage gradient between the center of the system and remote earth, reducing the resultant impedance of the ground system. This would in turn lower the voltage rise at the earth terminal, thereby providing better protection.

5.8 SIZE OF GROUNDING CONDUCTORS

Before describing practical grounding systems it will be worthwhile to determine in a general way the size of conductor needed for grounding. It appears that in existing grounding practice the conductor is often much larger than necessary.

Lightning-current surges are of short duration, and consequently the heat produced in conductors by the current is limited. The amount of energy in lightning strokes is considered in Chapter 3.

The required conductor size of any material depends upon the energy needed to raise it to its melting point. The energy depends upon the crest current and its decay time. An expression relating the current magnitude to the conductor size is

$$I_{\text{crest}} = KA/T \tag{2}$$

where K is a constant for the material of the conductor, A is its cross-sectional area in square millimeters, and T is the period of time in microseconds from the crest value of the current to half the crest value.

The values of K for several metals are given in Table 5.3.

TABLE 5.3 VALUES OF K FOR SEVERAL METALS

Metal	K
Copper	320,000
Aluminum	200,000
Bronze	200,000
Steel	220,000

The formula (equation 2) for permissible crest current versus conductor size for three values of T is plotted in Fig. 5.9. The values for T cover the normal range of wave shapes that occur. As the figure indicates, No. 10 AWG copper wire will withstand a crest current of 250 kA, a value that is not

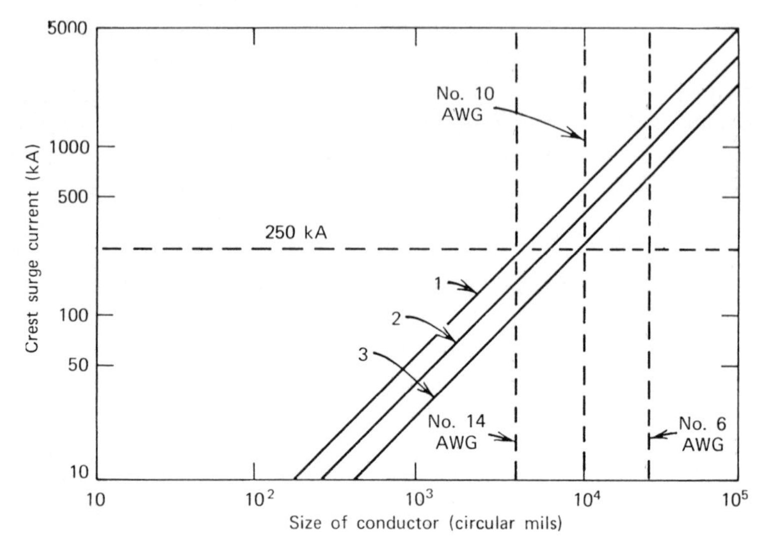

Fig. 5.9. *Size of conductor required for crest-current values of three steep-front wave shapes. Curve 1, 9-μsec tail; curve 2, 18-μsec tail; curve 3, 40-μsec tail. The 250-kA crest-current value is not exceeded in 99.99% of all strokes.*

exceeded in 99.99% of all lightning strokes. In practice a conductor of No. 6 AWG wire is recommended. Wire of this size provides a large safety factor and has adequate mechanical strength for most applications. Of course, ground rods must have sufficient radius to permit driving, and some bonding conductors or buses must be sufficiently large for bolted or brazed joints. However, in power-system practice, where power follow current might continue for several seconds, larger conductors are required. A duration of 4 sec is assumed in the IEEE recommendation.[8] The suggested sizes for copper conductors, where maximum temperatures of 450°C are allowed for brazed joints and 250°C for bolted joints, are listed in Table 5.4.

TABLE 5.4 SUGGESTED SIZES FOR COPPER CONDUCTORS[a]

Fault Current (A)	Conductor Size
Up to 5000	No. 2/0 AWG
5001–10,000[b]	250 MCM
10,000–25,000[b]	500 MCM
25,001	1000 MCM

[a] Data from Mukhedkar and Demers.[9]
[b] Minimum and maximum values.

5.9 EFFECTS OF SHARP BENDS

Bends in lightning conductors are undesirable, but a right-angled bend can be tolerated if firmly anchored. It has been calculated[10] that an extreme value of lightning current, 200 kA, passing through a right-angled bend will tend to straighten that bend with a force of 5500 ft-lb. This force can be withstood by firmly clamping the conductor in place, provided the radius of curvature of the bend is greater than 1 cm. However, if there is a reentrant loop in the conductor, as when it is doubled back under a parapet, the inductance of the loop can be sufficient to cause an arc across the loop, in addition to the mechanical forces on the bends.

5.10 CORROSION

Care must be taken in joining dissimilar metal conductors in the earth or in the open where moisture is present. In connecting a copper grounding system to a galvanized-steel tower the connection should be made above ground, protected from moisture, and accessible for inspection. If it is necessary to bond the aluminum sheath of a coaxial line to a galvanized-steel mast, this should be done with galvanized-steel straps. The zinc of the galvanizing will corrode first, rather than the aluminum, and the straps can be replaced periodically. Galvanized-mild-steel wire or rods are satisfactory earth electrodes, but if joined to a copper bus, inside or outside a building, the joint should be above ground and kept free of moisture.

At a junction of two dissimilar metals electrolytic action will take place, and the anodic terminal of the pair will corrode. The rate of corrosion will depend on the difference between the inherent potentials of the two metals. In the galvanic table of several metals (Table 5.5) the potential drop between any pair of metals in contact is the difference between their individual values in the table. For example, if there is galvanized-steel pipe buried near a copper-electrode ground system, it will assume a potential of 0.8 V negative with respect to the copper electrodes (this value will be 0.6 V negative if the pipe is of ungalvanized steel). This would result in depletion of the steel through galvanic action. The copper would be relatively unharmed, but it would attract a coating of salts due to the electrolytic action.

To prevent such erosion of steel pipe there are three alternatives: (a) to insulate the steel pipe against galvanic action, (b) to use steel ground electrodes instead of copper, and (c) to install "sacrificial" electrodes to attract the galvanic action away from the steel. The second method is the most economical and practical. In soil where there is likely to be erosion of

TABLE 5.5 GALVANIC POTENTIALS OF METALS

Metal	Relative Potential (V)
(*Corroded End, Anodic End*)	
Magnesium and its alloys	−1.6
Zinc diecasting	−1.1
Zinc plating on steel	−1.05
Chromate-passivated and galvanized iron	−1.05
Cadmium and cadmium plating on steel	−0.80
Wrought aluminum	−0.75
Non-corrosion-resisting steel or iron	−0.70
Cast iron	−0.70
Duralumin	−0.60
Lead	−0.55
Lead–silver solder (2.5% silver)	−0.50
Terne plate	−0.50
Tin plate	−0.50
Tin–lead solders	−0.50
Chromium plating, 0.005 in. on steel	−0.50
Chromium plating, 0.003 in. on nickel-plated steel	−0.45
Corrosion-resisting steel (12% chromium)	−0.45
Tin-plating on steel	−0.45
High-chromium steel, 18/2	−0.35
High-chromium steel, 18/8	−0.20
Brasses	−0.25
Copper and its alloys	−0.25
Nickel–copper alloys	−0.25
Silver solder	−0.20
Nickel plating on steel	−0.15
Silver and silver plating on copper	0.0
Titanium	0.0
Rhodium plating on silver-plated copper	+0.05
Carbon (colloidal graphite in acetone, evaporated to dryness)	+0.10
Gold	+0.15
Platinum	+0.15
(*Protected End, Cathodic End*)	

galvanized steel the use of sacrificial anodes would be justified in large systems. An elaborate system using magnesium rods as the sacrificial anodes has been described by Ficchi.[10] The ground rods are 40 ft long and made of steel, and the magnesium anodes are 10 ft long. The magnesium-anode holes were packed with salts (bentonite, sodium sulfate, and gypsum) to make good electrical contact with the soil. The resistance of each set of anodes and of each set of steel rods was measured, and the results are shown in Table 5.6. The earth at this location had soil, sand, and gravel in the upper layer and gravel, rubble, and silt in the lower layer. The sacrificial anodes in this case would protect the steel rods from erosion due to any dissimilar buried metals in the vicinity and as a side benefit would protect the buried steelwork of nearby structures from electrolytic action also. The joining of interconnecting cables to rods can be done by brazing, welding or the exothermic welding known as the Cadweld process, or by clamp-type fittings. Such joints should be sealed against moisture, and main connections to the earth-electrode system should be made demountable for test measurements.

TABLE 5.6 EARTH RESISTANCE FOR VARIOUS GROUND-ROD MATERIALS[a, b]

	Resistance (Ω)		
Location	Magnesium	Steel	Magnesium and Steel
1	4.2	3.3	3.0
2	5.8	3.3	3.1
3	2.8	2.0	2.0
4	6.25	3.0	2.8

[a] Data from Ficchi.[10]
[b] Thirteen rods and six anodes were used in each case.

5.11 SALTING OF GROUNDING ELECTRODES

In certain situations it might be justifiable and/or economical to lower the ground resistance by salting the soil near the electrodes with ordinary salt (sodium chloride). Such situations would occur where the soil resistivity is high or where the space and/or depth for electrodes is limited. Reductions in ground-system resistance by factors of 50 or more can be achieved by heavy salting, that is, by mixing in about 4 lb of salt per 100 lb of soil fill. Its action is to enlarge the effective radius of the electrode. If salt is mixed into the fill around a ground rod, out to a radius of 5 ft, the effective conducting radius of the rod is far in excess of 5 ft due to seepage of the salt solution.

The effect of salting on a ground electrode over a period of time is illustrated[10] in Fig. 5.10. The salting method requires attention and periodic replenishing. The ground-electrode connection should be made accessible so that the normal connection to it can be removed and its resistance to earth measured periodically. Ideally the salt in solution should be replenished just prior to the thunderstorm season, each year, unless the annual rainfall is too low to cause much seeping away of the salt solution.

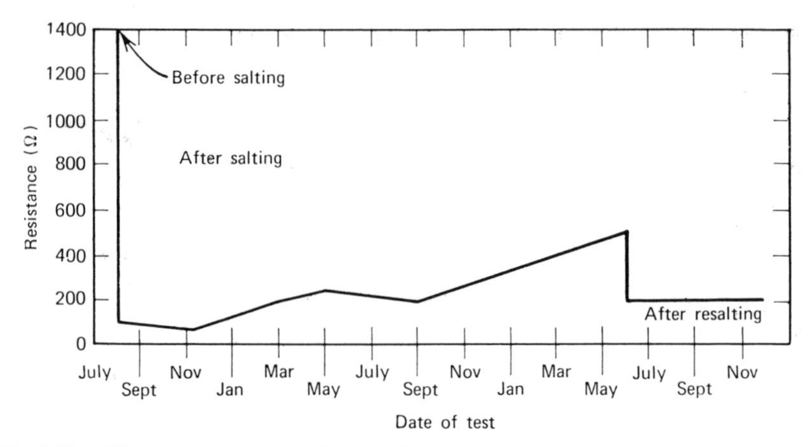

Fig. 5.10. *Changes in resistance of a ground connection in response to the presence of salt over a considerable period. After Ficchi.*[10]

Another beneficial effect of salting is that it maintains a low resistance value in the soil at lower temperatures than in unsalted soil. In plain soil the resistivity rises sharply when the moisture freezes. In salted soil the resistivity does not rise substantially until a temperature of $-10°C$ is reached, depending of course, on the density of the salt solution.

With salting there is likely to be more rapid corrosion of the earth electrode, so it must be examined yearly and replaced when necessary.

5.12 METHODS OF INSTALLATION

The method of installing ground electrodes will depend on the kind of soil that is present and the size of the ground system. If the soil is free from rock, ground rods 6 to 10 ft long can be driven in with a sledgehammer. For longer rods a chuck can be attached to the rod at working level, and a sliding hammer that encircles the rod is used for driving. It is also possible to use jointed rods and add threaded sections to screw into a coupling on the lower rod section, where deep penetration is required. For driving purposes a stud is screwed into the coupling to prevent thread damage. Where occasional rock occurs to obstruct a ground rod, a rock drill would have to be used to clear the way. In very large ground-rod systems a power hammer could be more economical than manual methods. For joining ground rods by an interconnecting wire conductor the recommended methods are by brazing or the Cadweld exothermic welding process. Where the ground conductors from the building tower or other structure to be protected are connected to the buried-earth-electrode system, a detachable connection should be made by clamps, bolts, and the like. This connection should be left exposed and accessible; and if different metals are joined, it should be sealed against moisture.

Ground rods are made in various diameters and lengths, the diameter increasing with length so as to provide the mechanical strength for driving. The range of diameters is $\frac{3}{8}$, $\frac{1}{2}$, $\frac{5}{8}$, $\frac{3}{4}$, and 1 in. Lengths vary from 5 to 40 ft, but usual lengths are 6, 8, 10, 12, and 16 ft, with diameters of $\frac{1}{2}$, $\frac{5}{8}$, and $\frac{3}{4}$ in. Rods are made of copper, steel, or copperclad steel. The latter type combines mechanical strength with good resistance to corrosion.

Horizontal ground conductors are buried in trenches either manually or by a plough, depending on the extent of the system and the nature of the soil. Where rock is encountered, drilling and blasting may be done if it is necessary to hide the conductor from view. Where a short length of exposed conductor can be tolerated, it would be cheaper to leave it exposed. Furthermore, the wire can be routed around rocks or trees, as it need not run in a straight line. The path of least resistance is usually the path of least electrical resistance also. Full advantage should be taken of any low-lying moist or wet areas in burying the conductor. The size of conductor should be at least No. 6 AWG wire (0.4 cm) for reasons of mechanical strength, and the metal can be copper, galvanized steel, or copperclad steel. The impedance of several buried radial wires is lower than one long wire of the same total length. Normally lengths greater than 150 ft should be avoided, except for the long "counterpoise" buried along the length of a power-transmission line, where each tower effectively "sees" two buried radials leading to the two adjacent towers; and the towers are in parallel to a lightning voltage striking one of them.

REFERENCES

1. A. L. Kinyon, "Earth Resistivity Measurements for Grounding Grids," *AIEE Trans.*, December 1961.
2. *Guide for Measuring Ground Resistance and Potential Gradients in the Earth,* AIEE Standard 81, May 1962.

3. G. F. Tagg, *Earth Resistances,* Newnes, London, 1964.
4. R. H. Golde, "The Lightning Conductor," *J. Franklin Inst.,* **283,** No. 6, June 1967.
5. A. B. Wood, "Lightning and the Transmission Engineer," *Electronics and Power,* **15,** June 1969.
6. L. V. Bewley et al., "Fixing Counterpoise Length," *AIEE Lightning Reference Book,* 1935.
7. J. R. Eaton, "Impulse Characteristics of Electrical Connections to the Earth," *General Electric Rev.,* October 1944.
8. *Guide for Safety in AC Substation Grounding,* AIEE Publ. No. 80.
9. D. Mukhedkar and G. Demers, "Ground Grid Design," *EIC Eng. Journal,* February 1968.
10. R. F. Ficchi, *Electrical Interference,* Hayden, New York, 1964.
11. P. L. Bellaschi et al. "Impulse and 60-cycle Characteristics of Driven Grounds," *AIEE Transactions*, Vol. 61, 1942.

BIBLIOGRAPHY

E. D. Sunde, *Earth Conduction Effects in Transmission Systems,* D. Van Nostrand, Princeton, N.J., 1949.

CHAPTER 6

Safety of Life

6.1 THE NATURE OF THE DANGER

It is well known that injury and death can be caused by lightning. There is a greater risk of this in the equatorial latitudes, where thunderstorms occur with greater frequency than they do in the temperate zones. The degree of injury inflicted depends on the particular circumstances of each incident, and the value of current to produce lethal results varies with the parts of the body affected and the physical condition of the individual. Accordingly it is not possible to draw up absolute rules for personal protection. However, it is possible to set down rules for avoiding positions of danger by being far enough removed from objects likely to be struck, by being bypassed while in a building, or by being sufficiently well insulated.

The danger to human or animal life is almost always from indirect contact with the electric potential from a lightning stroke, because it would be rare that a person standing, even in an open area, would happen to be the point of strike. However, for this reason and others a person would be prudent to adopt a prone position in open country when an apparent thundercloud is overhead.

The sources of danger from indirect contact with lightning can be described as follows:

1. Lightning striking a structure or tree can take a parallel path to earth through a person who happens to be close to the stricken object.
2. The intense electric field from a stroke near a person can induce sufficient current in the body to cause death.

3. Lightning that terminates on the earth can set up a high potential gradient over the ground surface in an outward direction from the point or object struck. A person near the point of strike who is touching the ground with two of his body members spaced apart (i.e., two feet or a foot and a hand) is endangered by the potential drop between the two points of contact. The potential drop could also be developed in a vertical element carrying a lightning discharge, such as a wall, post, or tree, and therefore across any person touching it at two separated points. This is distinct from the parallel-path strike defined in item 1.

6.2 CURRENTS DANGEROUS TO LIFE

The human being is a variable electrical impedance depending on age, weight, physical condition, dryness of the skin, and posture of the body at the time of electrical shock. From experimental data and observations it has been found on the average that the body is affected by electric current as shown in the chart[1] of Fig. 6.1, at power-supply frequency, provided the current is flowing through the chest region. The sensation of shock becomes painful when the current flow through the body exceeds 10 mA, and at a slightly higher value the patient is unable to "let go" of any electrode he might be holding. Breathing becomes difficult above 20 mA, and muscular paralysis begins at about 40 mA. Between 100 and 200 mA the current can be lethal if sustained for more than a fraction of a second. This is the current range that forces the heart into ventricular fibrillation. Current values higher than 200 mA can block the heart action and make the person unconscious, but will not necessarily be fatal if resuscitation is carried out immediately.

The magnitude of the applied voltage necessary to produce dangerous current values depends on the resistance of the body. This varies between wide limits. Between hand and foot, for example, assuming good electrical contact, the resistance is about 500 Ω, excluding skin resistance. The skin resistance varies from about 1000 Ω/cm^2 for wet skin to about 3×10^5 Ω-cm^2 for dry skin and even higher values on hands toughened by manual work. However, at voltages above about 240 V the skin is punctured, often inflicting deep burns and leaving only the internal impedance of the body to limit the current.[2]

For protection purposes a value of 500 Ω is commonly assumed for the body resistance between major extremities; and a figure of 1500 Ω for the resistance between the perspiring hands of a worker.

In Fig. 6.2 are shown the currents that would pass through the body from a 120-V circuit for several values of body resistance, from a minimum of

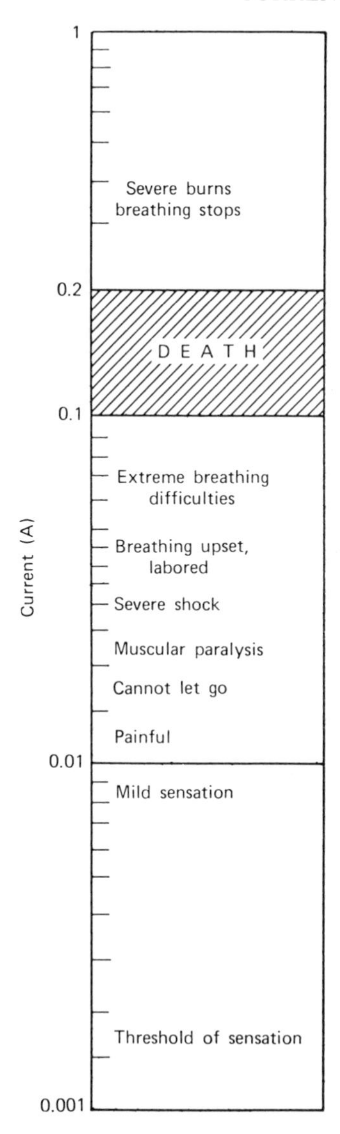

Fig. 6.1. *Physiological effects of electric currents. From Ref. 1.*

500 to a value of 20,000 Ω. It includes the let-go threshold for men and the range of current magnitude and duration that can electrocute adults.

At power frequency the impedance of the body is mainly resistive, but above 1000 Hz it exhibits reactance also due to the body's cellular structure and looks like a capacity in parallel with resistance. It is believed that very

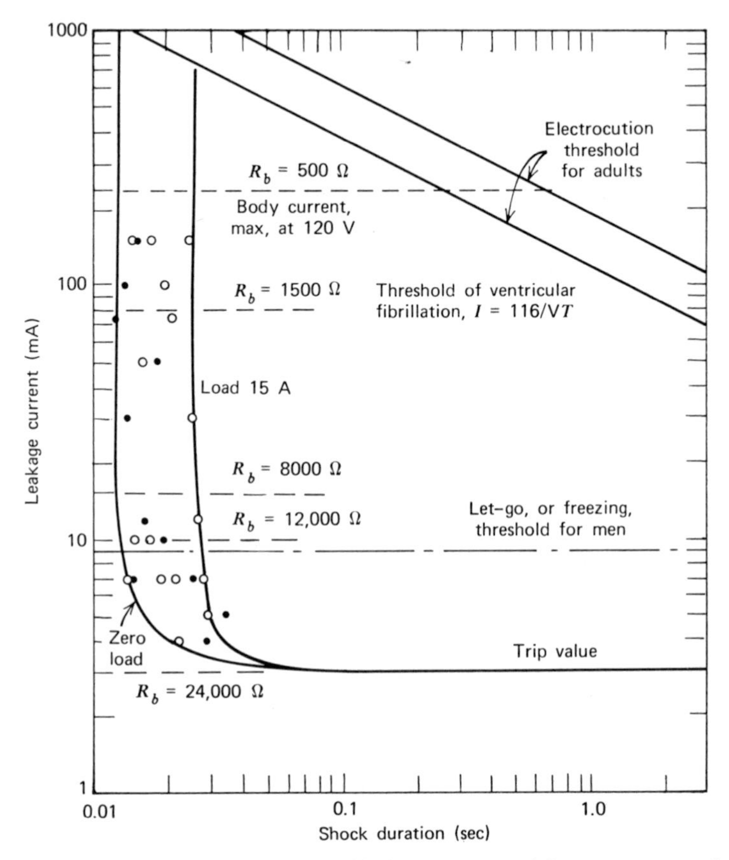

Fig. 6.2. *Body current for several values of body resistance and dangerous current-duration values. The curves starting down vertically from the left are the trip-current values of a ground-fault interrupter. After Dalziel and Lee.*[3]

high frequencies, above 100 to 200 kHz, burns are the main effects produced. The curves of Fig. 6.3 show the variation in let-go current with frequency for male subjects. The threshold rises rapidly for frequencies above 1000 Hz.

The smallest current that can cause a person to react involuntarily is called the reaction current. Because such movements can result in further injury, efforts have been made to limit leakage currents from electrical appliances. In the United States standards of 0.5 mA for two-wire portable devices and 0.75 mA for heavy, movable, cord-connected appliances like freezers and air conditioners have been established.

An effective protective device[4] against leakage currents to ground is the ground-fault interrupter. A sketch of its circuit is shown in Fig. 6.4, and its

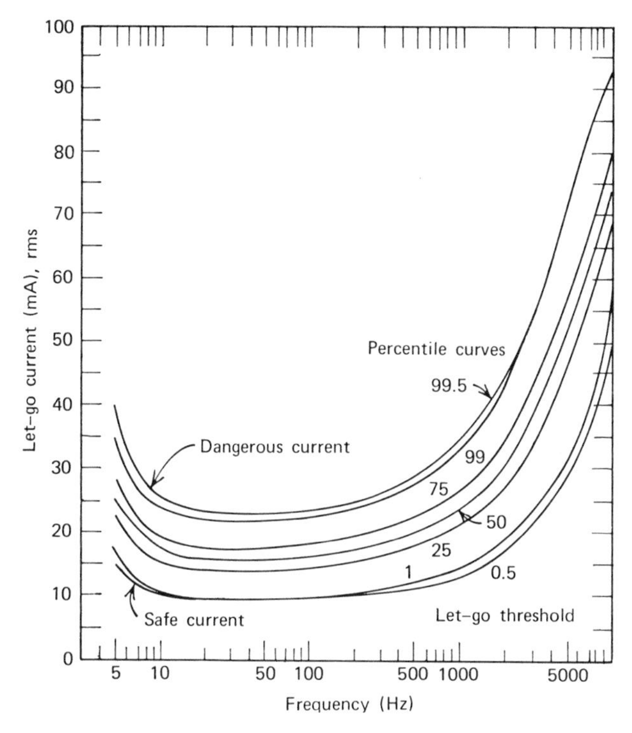

Fig. 6.3. *Effect of frequency on let-go current for men. After Dalziel and Lee.*[3]

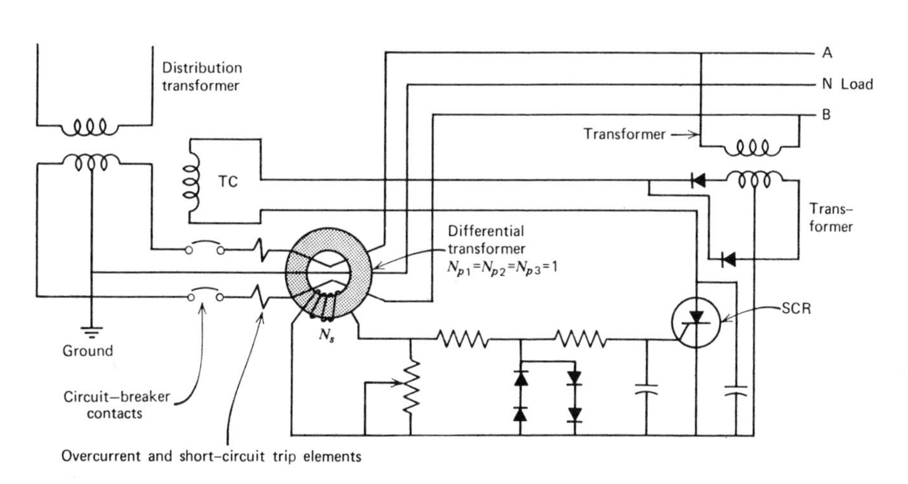

Fig. 6.4. *Solid-state ground-fault interrupter. The device has a sensitivity of 5 mA, which compares favorably with earlier types for 25 to 30 mA. After Dalziel.*[2]

performance is illustrated by the curves on the left side of Fig. 6.2. The differential transformer in the device detects small imbalance currents between each side of the power circuit and trips the circuit when they occur.

The upper curves in Fig. 6.2 indicate the current and duration values that produce ventricular fibrillation. It will be seen that 500 mA for 0.1 sec or 250 mA for 0.4 sec could cause fibrillation.

The term "ventricular fibrillation" means that the ventricular muscles of the heart are thrown into irregular movement. The normal heart has self-triggering tissue that periodically generates electrical pulses at 70 to 80 pulses per second. These pulses are eventually transmitted to the muscles of the ventricles, the main pumping muscles.

Fibrillation interrupts the regular pumping of blood, reducing the blood supply to the lungs and brain, and to other parts of the body. As a result of being deprived of oxygenated blood the brain may die within a few minutes or suffer permanent damage even if the victim is revived. Fibrillation can rarely be stopped except by the use of an external device, the defibrillator, which applies short strong shocks to restore rhythmic pulsing to the heart.

Currents above the fibrillation value may cause cardiac arrest, breathing stoppage, damage to the nervous system, and serious burns. However, if the victim is fortunate enough to be resuscitated quickly, he is less likely to suffer permanent injury than a fibrillation case.

In setting a limit of safety for the human a 2-A current for a short-duration shock of 100 msec may be used as a reference. A formula developed by Dalziel and Lee[3] from experimental and statistical data to determine the fibrillation current value is of the form $I = K/\sqrt{T}$ milliamperes. The quantity T is the time in seconds. The factor K has a value that is predicated on 95% of cases and an average weight of 50 kg; K is given a minimum value of 116 and a maximum of 185 in the above empirical formula for shock durations in the range of 8 msec to 5 sec.

The small currents we are discussing here, usually caused by faulty appliances or cords, are not the result of lightning, of course, but it is necessary to know the magnitude of lethal currents in order to consider protection against currents developed in the earth or elsewhere by lightning.

6.3 DANGEROUS VOLTAGE GRADIENT

From the allowable limits of body current one can calculate[5] the limiting voltage gradient along the ground that a person, standing, can tolerate.

In considering what voltage gradients are dangerous to human life, there is difficulty in surmising the posture of the body at the time of the lightning discharge. For a person walking or standing with feet apart an estimate can be made of the current that would pass through the body for a given voltage gradient along the ground. Let us say that a tower whose footing is equivalent to a hemisphere of some radius is discharging a lightning current I to earth. The current density spreading out radially from the tower base can be expressed as $I/2\pi x^2$ at distance x from the center of the tower base. If the soil resistivity is ρ ohm-meters, the current density will produce an electrical field potential, $E = \rho i$ volts, or field strength, $e = \rho I/2\pi x^2$ volts per meter. If the radius of the equivalent hemisphere of the tower footing is called B, the field strength will be a maximum at radius B, and its value is $e_B = \rho I/2\pi B^2$ volts per meter. If the discharge current I is 1000 A, ρ is 100 Ω-m, and B is 2 m, then $e_{max} = (100 \times 1000)/2\pi(4) \simeq 4000$ V/m.

Let us now assume that someone is standing at a distance x from the tower, with his feet separated by a distance s. The area of the earth under each foot is given an equivalent radius b of 7 cm, and its resistance can be expressed as $r = \rho/2\pi b$. The voltage drop between the two feet can be equated to the field strength times the distance between them, that is, $2ri_s = es$, where i_s would be the current going through the body. This does not allow for insulation by footwear nor for a relatively high skin resistance if the sole of the foot is dry. From the foregoing equation

$$i_s = \frac{es}{2r} = \frac{\pi bes}{\rho} = \frac{sb}{2x^2} I$$

at distance x. If we give the step distance s a value of 0.5 m, use the previous values for ρ and I, and assume the person is 5 m from the tower, we find that

$$i_s = \frac{0.5(0.07 \text{ m})(1000 \text{ A})}{2(5 \text{ m})^2} = 0.70 \text{ A}.$$

As will be recalled from Section 6.2, this could be a lethal value if it lasted for more than a tenth of a second.

The foregoing development assumes that the person has his feet separated along a radial line from the source of the voltage. If the feet separation were at right angles to such a radial line, there would be little or no voltage gradient between them, and the person would be safe, due to the element of chance. The "step" voltage due to voltage gradient along the ground and the "touch" voltage due to the vertical gradient along a tower or structure are shown graphically in Fig. 6.5, and the current distribution and voltage drop in ground due to lightning stroke to a building are shown in Fig. 6.6.

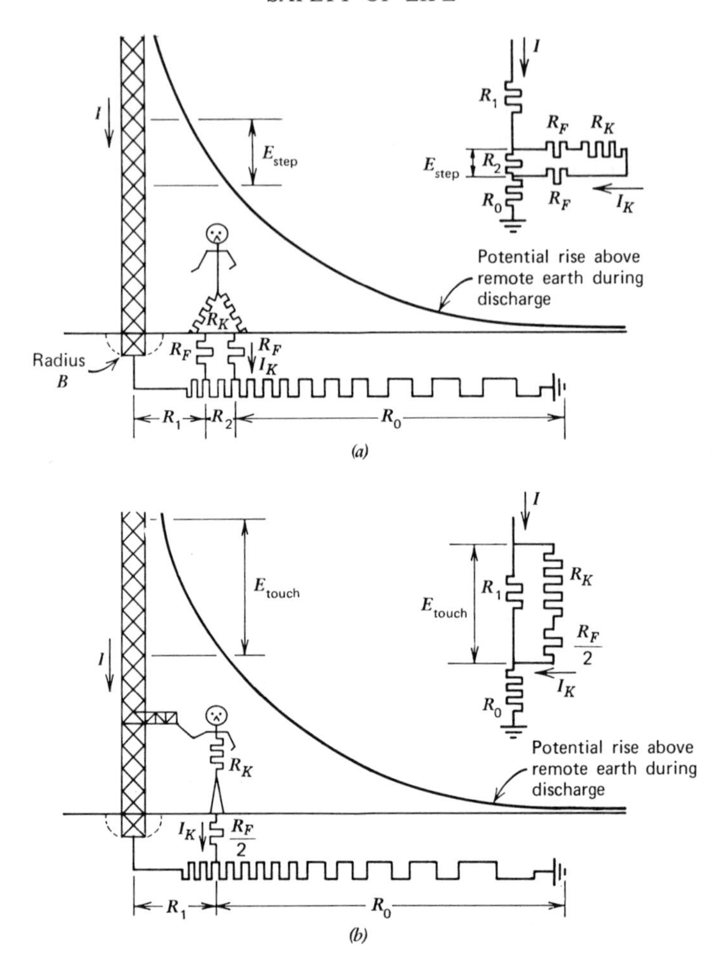

Fig. 6.5. *Step voltage (a) and touch voltage (b) at a grounded structure. From Ref. 6.*

Most of us are familiar with newspaper reports of injuries and fatalities caused by lightning. Deaths occur at beaches, on sports fields, golf courses, farm fields, at isolated dwellings, and so on. The injury can be caused by a direct stroke, by the intense electrical field from a stroke very close by, or by a voltage gradient along the ground caused by a nearby lightning stroke or along a vertical object the victim is touching.

In one of many such reports read by the author a soccer player was stunned when a nearby goalpost was struck by lightning. This illustrates the effect of a

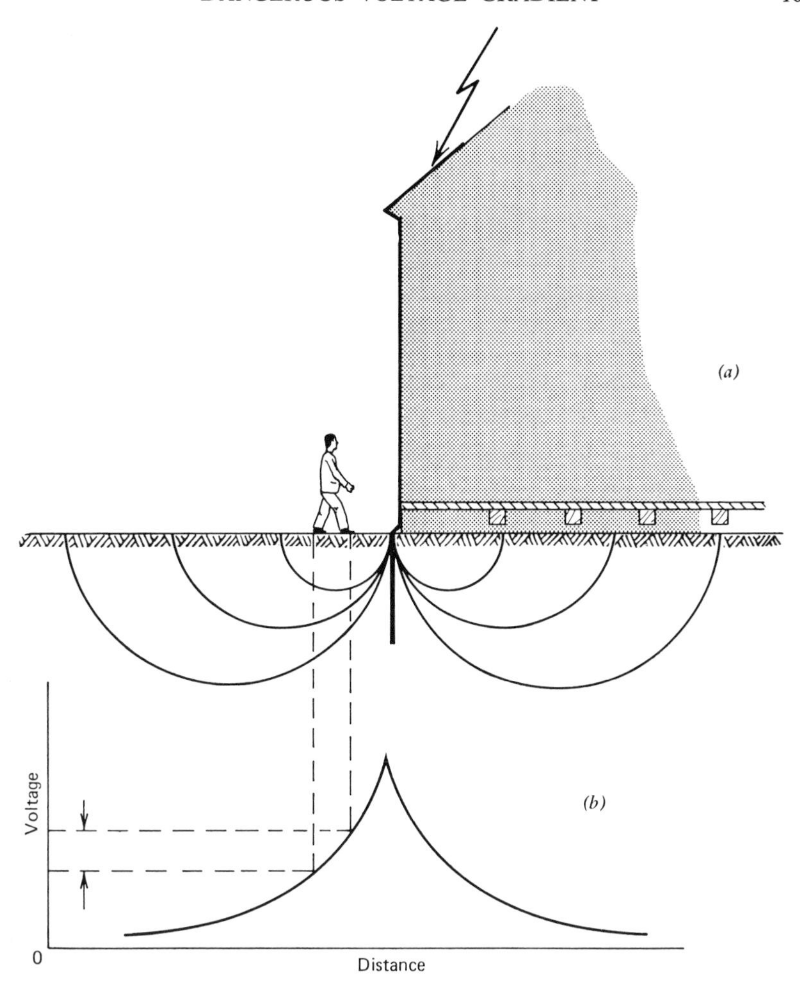

Fig. 6.6. *Current distribution (a) and voltage drop (b) in ground due to lightning stroke to a building. After Golde.*[7]

potential gradient along the ground extending outward from the goalpost. Another report described a boy riding a horse that was being led by his sister. The boy and the horse were killed by a lightning stroke, but the girl was unharmed. In this case the lead rope held by the girl obviously was sufficient insulation to save her, and the voltage gradient in the ground, perhaps combined with a fortunate position of her feet at the instant of the stroke, was not sufficient to shock her. In a farm accident four cows were killed in a field near a fencepost. Although no specific information was given, it seems likely that the lightning struck the fencepost, and the resultant voltage gradient along the earth killed the animals.

6.4 SAFETY OUTDOORS

In open areas, such as fields, meadows, parks, sports fields, golf courses, or beaches, a person standing or walking is the tallest terminal for the lightning and therefore risks being struck if there are thunderclouds overhead. The same remarks apply to a person standing in a lake or ocean (at a beach) and to a lesser extent to someone in a swimming pool. If one wishes to be safe under such conditions, he should get as low as possible, and if on the ground, he should keep the legs together and the arms close to the body so as to prevent a voltage difference developing between a limb and the body. One should avoid standing near an isolated tall object like a tree, pole, or mast; but within a grove of trees or bushes there would be comparative safety.

The safety rules to follow when a thunderstorm threatens are summarized below[8,9]

1. Do not work on fences, telephone or power lines, pipelines, or structural steel fabrication.
2. Do not use metal objects like fishing rods and golf clubs.
3. Do not handle flammable materials in open containers.
4. Stop tractor work and dismount, particularly when the tractor is pulling metal equipment. Tractors in open fields are often struck by lightning. However, a tractor with metal cab enclosing the operator would be safe.
5. Get out of the water and off small boats.
6. Stay in your automobile if you are traveling. Automobiles (with metal roofs) offer excellent lightning protection.
7. Seek shelter in buildings. If no buildings are available, seek protection in a cave, ditch, or canyon—or under head-high clumps of trees in open forest glades (but not near an isolated tree).
8. When there is no shelter, avoid the highest object in the area. If only isolated trees are nearby, it is best to crouch in the open, keeping away from isolated trees a distance at least equal to the height of the trees.
9. Avoid hilltops, open spaces, wire fences, metal clotheslines, exposed sheds, and any elevated metallic objects.
10. When you feel the electrical charge—if your hair stands on end or your skin tingles—lightning may be about to strike you. Drop to the ground immediately.

6.5 SAFETY INDOORS

Most houses are relatively safe because there are often taller trees or a pole nearby to act as terminals. However, for houses in open areas the protection methods described in Section 8.12 should be used.

Office blocks and other commercial buildings that have higher buildings nearby or have metal framing or metal reinforcing in the walls are likely to be safe for personnel. However, if a building is of wood or masonry construction and is isolated from other taller objects, it should be protected as described in Section 8.2.

If a house has no lightning protection system, the occupants should observe the following rules during a thunderstorm:[8]

1. Stay away from open doorways or open windows.
2. Stay clear of metallic objects like radiators, stoves, metal fireplaces, water pipes, and sinks.
3. Avoid touching plug-in electrical appliances like refrigerators, radios, television sets, and washers, and do not use portable plug-in appliances like electric razors or hair dryers; avoid using the telephone.

Although electric lines and telephones have lightning arresters, the voltage on them could rise to dangerous values before the protective device bypasses the lightning voltage to earth.

6.6 SAFETY ON THE FARM

A farmhouse without tall trees or other tall objects nearby should be protected as described in Section 8.12, and the precautions for personal safety listed in Section 6.5 should be observed. Barns should be made safe for the livestock therein. The principles of protection are the same as those for a residential building and are described in Sections 8.12 and 8.13.

The development of a dangerous voltage gradient along the ground near the earth terminal of a discharge was referred to in Section 6.3. As shown by Fig. 6.6, which depicts the current flow and voltage gradient from a buried earth electrode, a person is safer indoors than near a grounded structure during thunderstorms. Safety measures can be installed, but at a cost. For example, wire mesh or galvanized wire fencing of large mesh can be buried under walkways and bonded to earth electrodes. This would equalize the potential between the feet of an individual walking over the earth screen. Incidentally, the earth screen would lower the resistance of the earth-electrode system and consequently would lower the voltage gradient over the surrounding area.

6.7 PROTECTION IN CABINS AND TENTS

Any shelter for humans, whether it be a cabin or tent, needs protection if there are no trees or other higher objects nearby. It needs an air terminal, down conductors, and an earth terminal. The air terminal should be one or more wires strung over the highest part of the roof. Two or more down conductors should lead to a buried earth system, which should consist of two or more ground rods or the equivalent in horizontal conductors. In the case of a tent it is important to encircle it with a buried ring conductor bonded to the buried electrodes. This equalizes the potential in the earth with respect to the occupants of the tent.

Specifically the elements of the grounding system for a tent would be a horizontal conducting wire strung just above the ridge pole, supported at each end. From the supports the wire should continue down to a ground rod at each end of the tent. For practical reasons the length of the rods can be limited to 4 ft. Then the tent should be encircled with a bare wire buried a few inches in the earth and wrapped around each ground rod. In this case, although the resistance of the earth system might not be as low as desired,

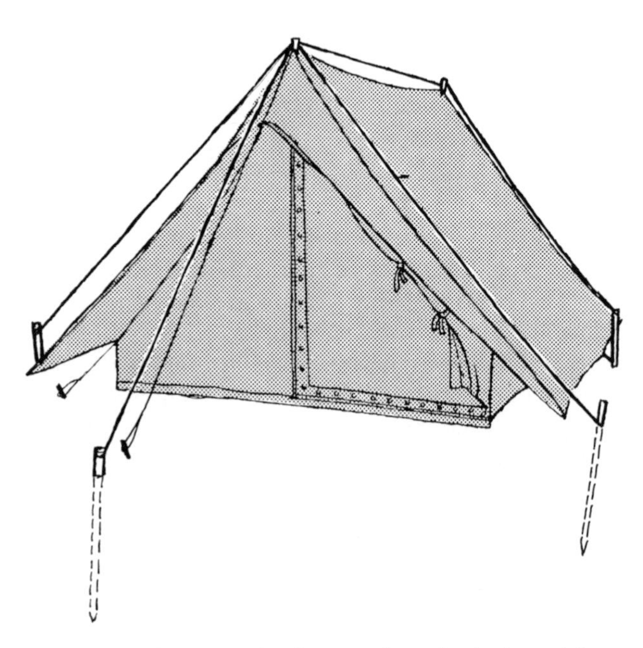

Fig. 6.7. *Protection of tent in isolated area with overhead wire and four ground rods (one ground rod is at the rear).*

the rise in potential of the earth-floor of the tent, from a lightning discharge to the grounding system, will not injure the occupants. The buried ring conductor will prevent a dangerous voltage gradient from developing across the tent floor.

A simple protection scheme for tents is sketched in Fig. 6.7. One should avoid having an isolated shelter on a rock base. If there is bedrock with a layer of soil on top, a horizontal buried wire about 50 ft long should be used in place of each rod. A new type of European tent[10] has conductors in its fabric so as to form a Faraday cage for lightning protection. Also a tent supported by metal tubing that passes over the top can provide excellent lightning protection when the lower ends of the tubing are connected to ground rods or a buried wire conductor.

A cabin with a roof that rises to a peak can be protected the same as a tent. For a flat (or slightly sloping) roof two wires at right angles should pass over the roof, crossing near the center of the roof and terminating on ground electrodes, four in all.

6.8 SAFETY IN VEHICLES

An automobile with a metal body and roof affords safety to the passengers during a thunderstorm. Under dry conditions the tires may insulate it from earth, so that both the metal body and occupants could be raised to a high voltage momentarily, but without detrimental effects. Under wet conditions any discharge to the car body would be drained to earth. At one time drag chains were used on trucks carrying flammable or explosive materials. However, it is generally considered that the chain does not provide a low resistance to earth under dry conditions, and it is not needed under wet conditions.

Although aircraft of the usual type with metal skin can suffer damage from lightning, they are rarely forced down by lightning. The kind of damage caused is burning of the metal skin, damage to projections like radomes, and disruption of electronic navigation and communications equipment, often by the magnetic effect of the lightning current discharge. In one case[8] an aircraft crash was attributed to lightning's igniting the reserve fuel supply. In general, however, occupants of metal aircraft will be unaware of a lightning strike.

Probably the most sensational lightning event of our times was the discharge that occurred across the Apollo 12 spacecraft just after liftoff. This was a case of an object originally at earth potential rising toward a charged cloud. A discharge is reported to have occurred while the vehicle was penetrating the cloud. The metal skin of the craft at one potential would induce a flash from a section of the cloud charged to a different potential. The effect of the flash or its electrical field was to render some electronic equipment inoperative by tripping circuit breakers. This incident led to a policy of avoiding thunderclouds in subsequent space flights.

6.9 PROTECTION ON TEMPORARY STRUCTURES

Workmen on such temporary structures as scaffolding and those installing the steel frames of structures should be protected if they are at the highest point in the vicinity. An elaborate buried-electrode system might not be justified or practical at the time, but connection should be made to one or two ground rods or other buried metal. For temporary grounds of this kind the salting of electrodes would be a practical method. Workmen stringing cables at substantial heights are vulnerable to lightning strokes under thunderstorm conditions, and therefore precautions should be taken to ground the cable, unless the men are electrically isolated from ground. In this regard an insulated suit has been developed to withstand a stress of 300,000 V. It is intended for men working on live high-voltage transmission lines.

Another type of safety suit made for live-line working on high-voltage transmission lines is made with a carbon-mesh lining (United States) and with a stainless-steel wire-mesh lining (Britain). These suits may be regarded as Faraday cages surrounding the workman. The worker is raised to the live conductor in a "chair" suspended by an insulating rope and can then work while his suit is in contact with a high-voltage conductor. Although these suits are intended for live-line working, they could be used for protection against lightning voltages by those working on tall structures during thunderstorm conditions.

REFERENCES

1. Tektronix Co., Serviscope No. 35, December 1965.
2. C. F. Dalziel, "Electric Shock Hazard," *IEEE Spectrum,* February 1972.
3. C. F. Dalziel and W. R. Lee, "Lethal Electric Currents," *IEEE Spectrum,* February 1969.
4. C. F. Dalziel, "Transistorized Ground Fault Interrupter Reduces Shock Hazard," *IEEE Spectrum,* January 1970.
5. R. Rudenberg, "Fundamental Considerations on Ground Currents," in *Grounding Principles and Practice,* AIEE (reprint of articles from *Electrical Engineering,* 1943–1944), June 1945.
6. *Measuring Ground Resistance and Potential Gradients in the Earth,* AIEE Bulletin No. 81, May 1962.

7. R. H. Golde, "A Plain Man's Guide to Lightning Protection," *Electronics and Power,* March 1969.

8. *Lightning,* ESSA/P1 660024, U.S. Department of Commerce, Environmental Science Services Administration, Washington, D.C., December 1966.

9. *Lightning Protection for the Farm,* Farmer's Bulletin No. 2136, U.S. Department of Commerce, Washington, D.C., Jan. 1968.

10. J. Wiesinger, "Tentes de protection contre le coups de foudre," *ASE Bulletin,* No. 59, 1968.

CHAPTER 7

Grounding of Communications Towers and Systems

7.1 GENERAL CONSIDERATIONS

Communications installations such as radiotelephone, FM or television, and microwave stations require lightning protection for several reasons. Usually they consist of a tall steel tower, situated on high ground, and an adjacent equipment building. In most cases the elevated ground is also rocky ground with a high resistivity. These towers are likely recipients of lightning strokes, and if the lightning discharge potential could not be diverted to earth, it would exert destructive effects on the equipment housed in the adjacent building, via the transmission line or cable that runs between tower and building. Power lines and communications circuits also reach the building on pole lines (usually), so they are subject to the lightning potential from the tower. Conversely any lightning strikes to the power or telephone lines at a distance can produce dangerous voltages at the building if it lacks a low-resistance ground system. The principal requirements for an effective grounding system may be summarized as follows:

1. A buried earth-electrode system centered on the tower to dissipate the lightning-stroke energy and thereby reduce the potential rise at the tower base.
2. A buried earth-electrode system at the building to supplement the main system in lowering the voltage surges on the incoming transmission lines and other circuits, and to equalize the earth potential between the building and tower.

3. Bonding together equipment panels, cabinets, and other building metal, and connecting it to the common ground system so as to prevent potentials arising between these components.

4. Suitable lightning gaps, or arresters, on power-line and other external circuits to divert dangerous potentials to the ground system.

7.2 KINDS OF DAMAGE

On the tower structure proper there will be one or more runs of coaxial transmission line from the antenna; power cables; and occasionally a communications circuit. Both conductors of the coaxial line will reach the metal of the mast for direct-current purposes at the upper end unless an amplifier intervenes between line and antenna. However, along its length down the mast the outer conductor would carry a proportion of the lightning discharge current, and its potential may differ from adjacent parts of the tower if the mechanical clamping does not also make good electrical contact. The result could be arcing to the coaxial line, with resulting damage.

In one case a 1000-ft television tower that had insulators in the upper section of its guy cables (to reduce the effect of each guy on the antenna radiation pattern) collapsed and fell during a lightning storm. The insulators, formed of nylon strands, had become somewhat spongy due to the ingress of moisture through a protective coating. It seems probable that the lightning discharge arced across the insulation, overheating it and causing it to rupture. The surge impedance of the tall tower, combined with a relatively high resistance to earth, would raise the tower potential to the point of causing arcing across the insulators. It is possible that had the tower had a low-resistance grounding system, the accident would not have happened. An additional recommendation would be to bypass each insulator with an enclosed spark gap and install a ground system at each guy anchor. This would provide several parallel paths to earth for lightning currents, thus reducing the potential rise on the tower, and would remove the risk of insulator failure. Grounding systems for towers are described in Section 7.3.

The nature of the damage that can be caused by a lightning discharge is varied. Thin conductors like those used in telephony can be melted by the temperature rise produced by the high lightning current. Furthermore, lightning hits on telephone lines that do not have a low-resistance ground terminal may enter the building and cause arcing between the telephone junction box and other metal in the building. Since this may char the interior wall in the process, the danger of fire is present. Externally such conductors should be strung below larger conductors, earth wires, or a grounded messenger wire.

Power lines terminating at the building, if not bypassed by protective gaps, or arresters, to a low-resistance ground system, can, if struck by lightning, cause a burnout of the step-down transformer, damage to switchgear, or fire.

If a coaxial transmission line or cable is installed without the provision of a low-resistance ground system, a high potential on the outer conductor due to a lightning strike can produce side flashes from the line to other metallic elements that are at ground potential. Furthermore, a flash between the outer and inner conductors can occur, possibly causing permanent damage to the transmission line. In severe cases a coaxial transmission line can be bent by the magnetic field of a high lightning current, and companion circuits for telephone, remote control, and the like can be destroyed by side flashes at the same time. A known example caused damage and physical distortion to the radio frequency transmission line and adjacent control cables at a radio broadcasting station that had four tall towers and an associated low-resistance ground system. The most likely explanation of the event is that lightning struck the power-supply line some distance away and flashed over to the coaxial transmission line at some point near the building. From the coaxial transmission line the surge impedance looking toward the towers would be relatively high due to the antenna matching circuits at the ends of the transmission line. This would permit a high potential to build up on the transmission line and produce arcing to other conductors and to the ground at supporting poles or other points along the line. The resulting high current would produce the magnetic force to distort the line, and the side flashes would melt the metal where the arc struck.

Caution should be observed with regard to "foreign" buried coaxial cables in close proximity to a buried earth-electrode system. There is a risk that the lightning-induced high potential on the buried system will cause electrical breakdown of the soil between the earth-electrode system and the buried coaxial line or other buried cables. The ensuing arc current would damage the buried conductors. The breakdown-voltage gradients of different soils are given in Table 4.4. Let us say that we are concerned with gravelly soil and that its approximate breakdown-voltage gradient is 15 kV/cm. Let us also assume that the resistance of the earth system is 20 Ω and that we want to protect against lightning-stroke currents of up to 150 kA. Then the potential arising at the ground-system terminal at the instant of strike is 150×20, or 3000 kV. So the soil would break down out to a radius of 200 cm, or 2 m. Accordingly any buried conductors should not be closer than 2 m to the buried earth electrodes.

If the buried conductors are part of the installation and are small, they should be run in metal conduit, but insulated from it. Larger coaxial lines may be buried, but the outer conductor should be bonded to the earth-electrode system. If the cable has a solid dielectric, a by-pass device, such as a gas-type arrester, may be connected between the inner and outer conductor so as to prevent breakdown of the dielectric.

On the other hand, if the grounding system is kept at a safe spacing above the buried conductors, it protects them by dissipating the lightning charge and reducing the potential rise in the earth.

7.3 TOWER GROUND SYSTEM

At most communications tower sites it would be difficult to achieve a ground-system resistance as low as 10 Ω because of the high resistivity of the soil material. However, if a moderately low resistance can be obtained, a satisfactory degree of protection can be realized if the buried earth electrodes are evenly distributed around the station so as to equalize the potential.

Beginning with the "air terminal," or topmost part, of the mast or tower, it is necessary to bolt or bond to the structure all such metallic fixtures as beacon lamps, antenna support mast or frame, and mounting frames of microwave reflectors. For a beacon lamp mounted at the top of the tower an air-terminal rod should be installed beside it, bolted to the tower structure and protruding 2 ft above the lamp. Coaxial transmission line and cable-carrying conduits should be bonded to the tower with metal straps every 50 ft.

It is unusual for a communications tower to be located on deep soil of loam and clay, but where this does happen the grounding problem is simple. For example, if the resistivity is 200 Ω-m, then eight ground rods driven 10 ft deep and spaced at least 10 ft apart would have a combined resistance of 10 Ω. This would be adequate, and the building ground bus, if connected to the rod-system, would protect the building contents.

In the more usual case where the surface soil is shallow and based on rock or gravel, one should take advantage of the foundation excavations in order to lower the resistance to earth. A "cage" of vertical wires (No. 6 AWG or larger) should be placed against the walls of the excavation, spaced apart by a distance equal to their depth and bonded together by a ring conductor of the same size wire at the rock surface. For example, in an excavation 6 ft square and 6 ft deep there would be a vertical conductor at each corner. If there is surface soil, buried radial conductors should fan out from the ring conductor of the cage at every 60 degrees to a length of up to 150 ft. These conductors may be routed around exposed boulders or left exposed for parts of their length where necessary. If the antenna mast is guyed, a cage should be installed in each guy-anchor excavation and connected to the lower end of the guy. If, on smaller towers, a metal guy anchor in the ground is used, it will also serve as a ground rod. In the case of a self-supporting tower with four legs supported on footings, cages in the foundation holes connected together would be roughly equivalent to a mast with one central foundation cage and three guy anchor foundation cages.

These cages are nearly equivalent for grounding purposes to a continuous cylinder of the same dimensions. The expression for the resistance to earth of such a cage is

$$R = \frac{\rho}{2\pi}\left(\log_e \frac{4l}{a} - 1\right)$$

where a is the radius of the equivalent cylinder expressed in the same units as l, the length or depth. In calculating the combined resistance of four such cages, corresponding to four tower legs, each cage may be regarded as a large-diameter rod, and the value for their combined resistance is obtainable from Fig. 4.13, which gives the ratio of the resistance of four rods to the resistance of one rod.

At the main connection points to a buried earth-electrode system at the tower, and at the building, the junction should be made detachable (i.e., by bolts or clamps) so that resistance measurements can be made before the final connection is made and at regular periods thereafter.

Details of a welded ground strap connection at a tower footing are shown in Fig. 7.1.

For the case of a guyed tower that has insulators in each guy near its point of attachment to the tower or near the very-high-frequency antenna, such as the 1000-ft tower mentioned in the preceding section, the insulators should be bypassed for lightning protection. This can be done with an enclosed protective spark gap connected across the insulator. The voltage rating of the gap could be low, say 1000 V, as it is only subjected to the VHF electrical field from the antenna.

It is unusual for equipment at a medium-wave radio broadcasting station to suffer much damage from a lightning strike to an antenna tower because there is usually a low-resistance ground system of buried radial wires. However, very large lightning-discharge currents, that jump across the usual ball gap or the more recent enclosed gap type of arrester at the tower base can build up a sufficient voltage drop to cause arcing and equipment damage. Furthermore, static voltages can build up on the antenna towers due to passing thunderstorm clouds, snow or hail, or dust particles carried by the wind. Sparkovers due to these sources do not normally cause damage, but they do cause momentary interruptions in service.

In regions subject to such conditions it is advisable to provide a low-impedance path to ground whenever possible, in order to reduce interrup-

Fig. 7.1. *Brazed ground strap connection to tower base steel at TV and FM transmitting station on high ground. Courtesy of the Canadian Broadcasting Corporation.*

tions. Two methods of doing this are illustrated in Fig. 7.2. Diagram (*a*) shows a π-circuit with coils L_1 and L_2 in the two shunt arms that provide a ground path for voltage surges. This arrangement is adaptable to low-impedance antennas. For high-impedance towers, where a T-circuit might be necessary for matching, as in diagram (*b*), a shunt coil L can be added to bypass voltage surges and a condenser C in parallel to form an antiresonant circuit of high impedance at the wanted signal frequency.

7.4 BUILDING GROUND SYSTEM

The purpose of the building ground system is to supplement the tower ground system and to provide a short route to ground for any lightning potentials that will be carried to the building by transmission, power, or communications lines. This also prevents a high potential difference from developing between the tower and building grounds. Power and communications lines

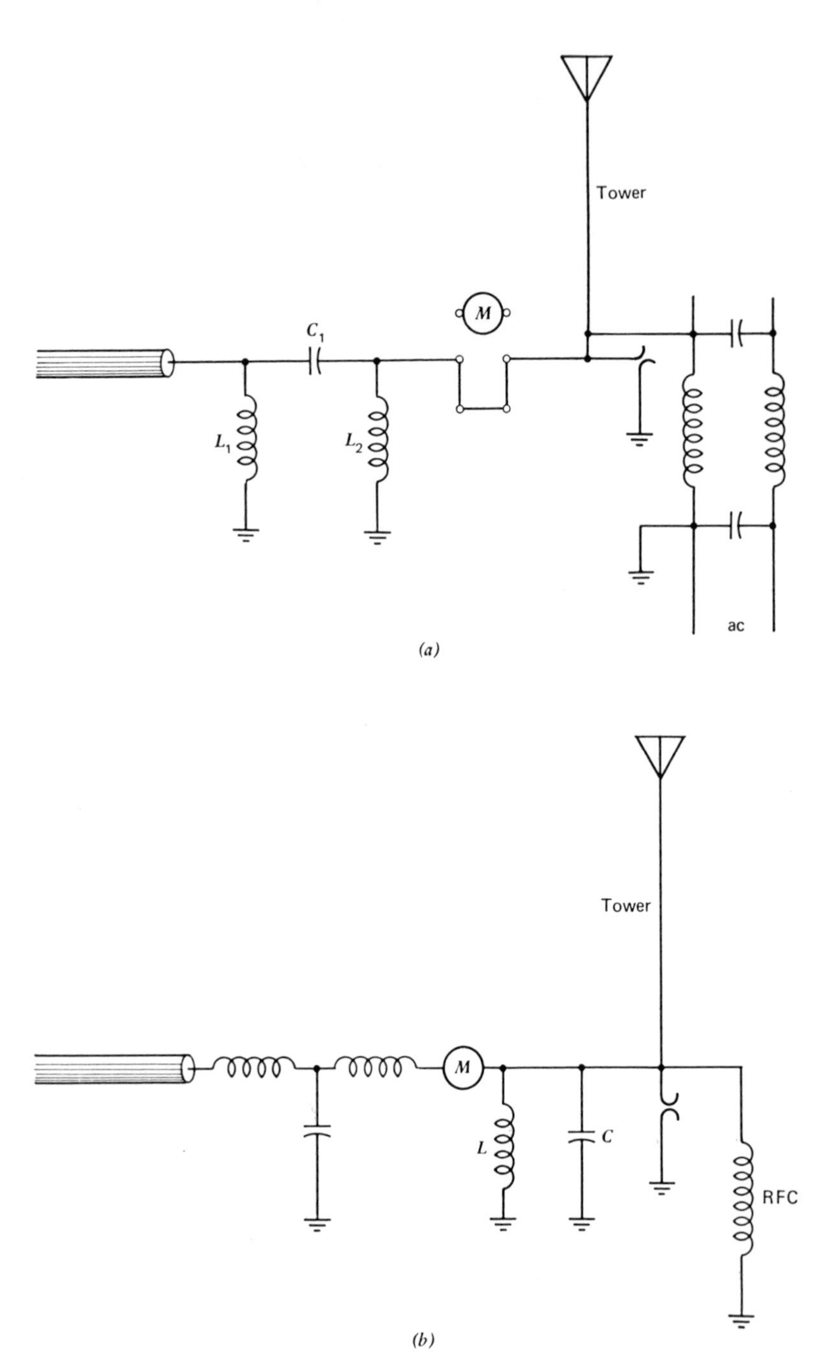

Fig. 7.2. *Methods of providing a low-impedance path to ground:* (a) *matching coils* L_1 *and* L_2 *provide lightning path to ground;* (b) *added coil L serves that purpose.*

are grounded via lightning arresters, whose function is to break down at overvoltage threshold values and permit the lightning energy to drain to the ground system. As such potentials could flash over to equipment racks or cabinets, these and all other metallic elements in the building must be bonded together and to a ground bus.

The building should be located within the protective zone of the tower—that is, within a radius equal to the overall height of the tower structure.

If the building has an excavation for foundations or basement, a cage should be installed as described for tower footings, but larger of course. The ring conductor interconnecting the vertical wires of the cage can also serve as the building ground bus and should therefore be larger, for bonding purposes (i.e., No. 2 AWG wire). If the building has no excavation, the encircling buried-ring bus should be installed. In both cases buried radial conductors should lead outward from the ring bus every 60 degrees to a length of up to 150 ft, but they should not intersect with the buried radials of the tower, as this would be an unnecessary duplication of contact with the earth. The building grounding system should be tied to the tower ground system by a short length of buried wire, No. 6 AWG or larger.

A sketch of a typical ground system is shown in Fig. 7.3. The estimated resistance to earth of such a system for a tower and building is given in Table 7.1 for several values of earth resistivity.

TABLE 7.1 APPROXIMATE RESISTANCE TO EARTH OF GROUND SYSTEM OF FIG. 7.4, FOR SEVERAL VALUES OF EARTH RESISTIVITY

Earth Resistivity (Ω-m)	Resistance to Earth of Ground System (Ω)
2000	15
5000	25
7500	35
9000	40

If a well is drilled on the premises for a water supply, it should be incorporated into the grounding system. If the well casing is metallic, a connection should be made to it with a suitable clamp. If only the supply pipe in the well is metallic, then, of course, the ground connection should be made to that. Where possible the resistance of the well pipe to earth should be measured prior to the installation of other buried earth electrodes because it might reduce the requirement for the remainder of the buried ground electrodes.

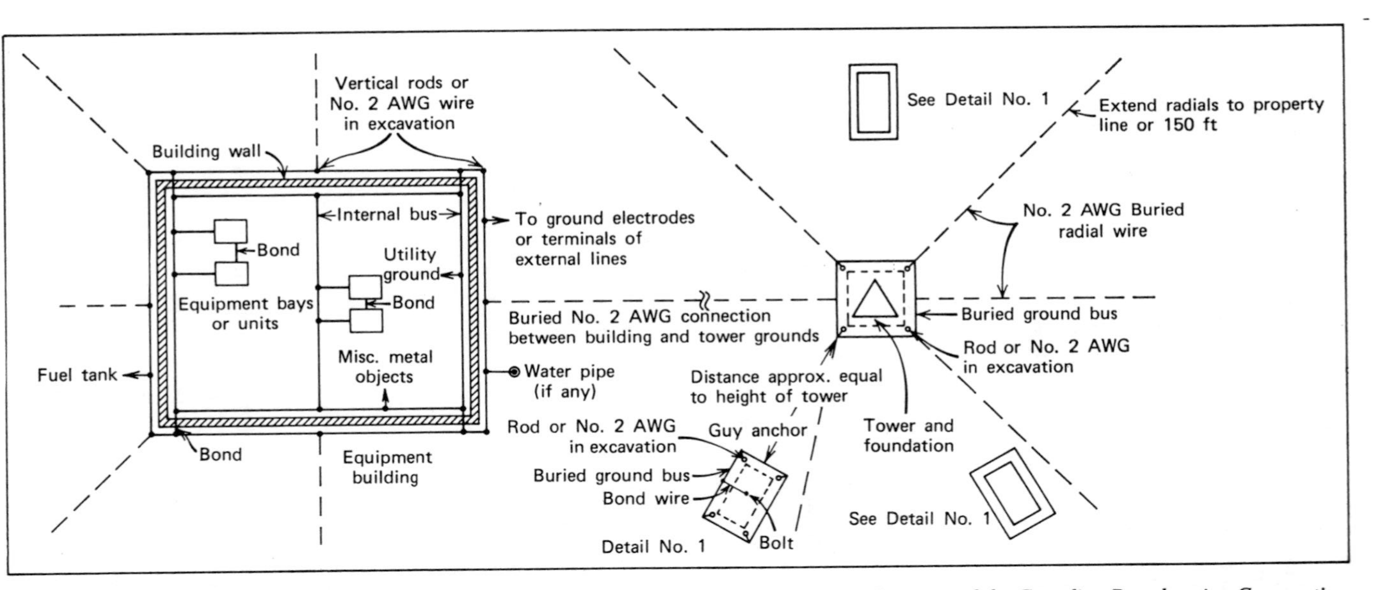

Fig. 7.3. *Schematic diagram of typical grounding system for TV, FM, or microwave station. Courtesy of the Canadian Broadcasting Corporation.*

Insulated horizontal runs of conductor connecting ground terminals should not be run in conduits. This arrangement would act like a section of coaxial transmission line to a current surge, presenting some value of reactance that would raise its impedance to earth.

An example of a grounding arrangement for a TV transmitting station building situated on a rocky hill is shown in Fig. 7.4A. The antenna tower, not shown in this diagram, has ground conductors in its foundation excavation, and there is a connection between tower and building via the copper coaxial transmission lines.

It will be seen in the figure that copper-strap grounding conductors were run across the building under the floor slab in order to lower the ground-system resistance somewhat and to equalize the potential across the building. The internal building grounding system further ensures that the potential will be equalized across the building. As a safety precaution for personnel who might have to be outside the building during thunderstorm conditions, a wire mesh 6 ft wide was laid around the building and between the building

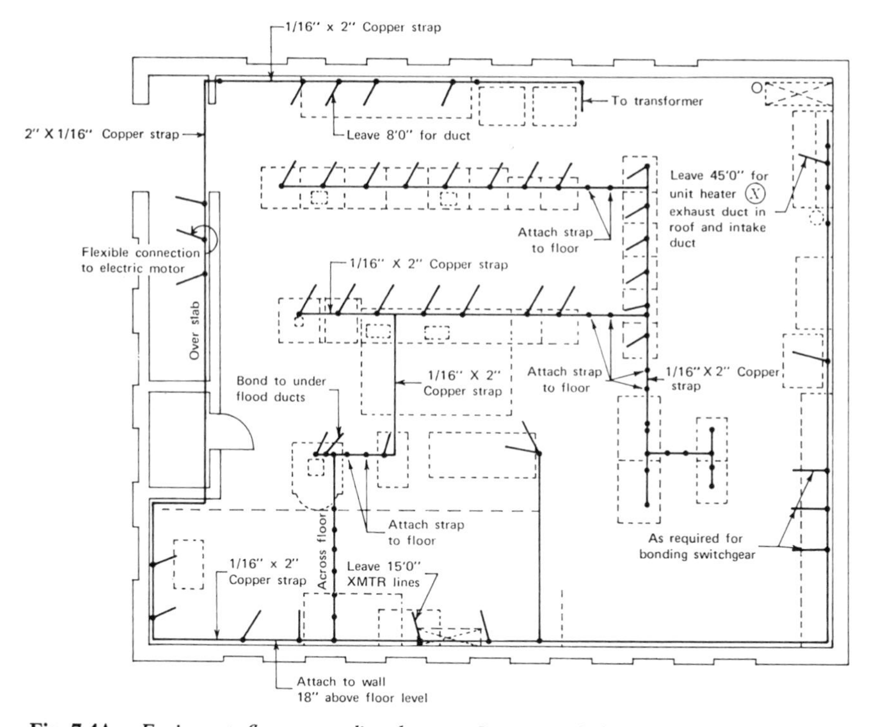

Fig. 7.4A. *Equipment floor grounding layout. Courtesy of the Canadian Broadcasting Corporation.*

and tower. The wire mesh is bonded to the ground conductors so that the whole system remains at the same potential. The measured earth resistivity at this location averaged 5000 Ω-m, and the grounding-system resistance measured 25 Ω, which was the average of measurements along three radial directions from the tower. This is a relatively high value in absolute terms, but is about all that can be done within reasonable cost. Furthermore, by arranging the ground conductors and wire mesh so as to ensure that the system maintains a common potential under lightning conditions a high degree of protection is achieved.

7.5 INTERNAL SYSTEM AND BONDING

As stated in preceding sections, it is necessary to bond together all metallic elements in the building and connect them to the ground system. A convenient method is to install a second ring bus round the inside of the building and bond it by short conductors to the outside buried-ring bus at each corner. Metal cabinets, panels, conduits, and the like, as well as circuit grounds, should be bonded to the building ground bus by short leads of No. 6 AWG wire. The building ground bus may be 1.5-in. copper strap for convenience and ease of attachment to the wall. To prevent ground loops in electronic circuits their individual "grounds" are often connected to a separate bus that is then connected to the buried earth system at only one point. A simple layout of internal bonding conductors is shown in Fig. 7.3, and a more elaborate system in Fig. 7.4B. Electrical panels and switches are usually connected to an equipment ground bus, and technical (electronic) equipment is connected to a "technical ground," which is an insulated conductor run inside a metal conduit. The incoming telephone and program circuits also use the technical ground. The ground buses for both systems are brought together at a main grounding terminal box where they can be joined or isolated by links. The isolation is useful in measurements and in tracing unwanted signals induced in ground conductors. At the main terminal box a connection is made to the buried earth-electrode system and/or metal water main. This connection is made demountable so that the resistance of the buried electrode system can be measured.

7.6 EXTERNAL CIRCUITS ENTERING BUILDING

It has been stated that transmission lines, communications circuits, power lines or cables, and the like should be taken to ground through their protective gaps, or arresters. The connection to the building ground system

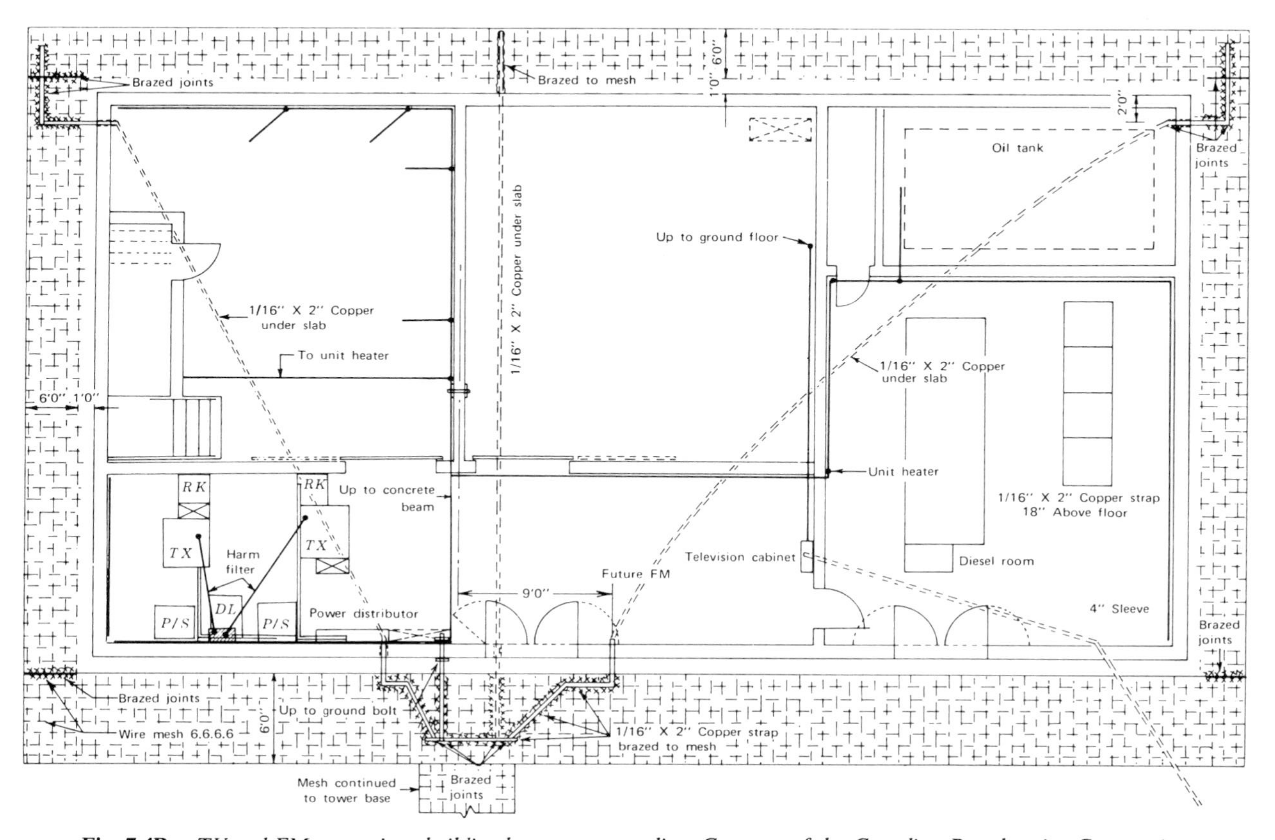

Fig. 7.4B. *TV and FM transmitter building basement grounding. Courtesy of the Canadian Broadcasting Corporation.*

should be made outside the building, so as not to risk any damage inside the building from a high current discharge.

7.7 SAFETY OF PERSONNEL

As explained in Chapter 6, means should be provided for bypassing lightning discharge currents so as to protect persons. In an equipment building where equipment bays or cabinets are distributed, the contact of these units with the floor and with the grounding system will keep the floor at the potential of the grounding system, thereby making it safe for personnel. The building ring ground bus will also maintain an equipotential on all sides of the building. Any metal framing or reinforcing in the walls should be bonded to the common ground system. Otherwise the recommended procedure is to attach a second ring conductor (e.g., one made of No. 6 AWG wire) to the upper wall, about 5 ft above the floor, and connect it to the main building ring bus at the four corners of the building. This will prevent a potential drop being applied across a person touching the wall during a lightning discharge, when the floor of the building would suddenly be raised to the ground-system potential.

7.8 PROTECTION OF COMMUNICATIONS AND OTHER LOW-VOLTAGE CIRCUITS

The sensitive electronic components of telephony, broadcasting, and other communications systems as well as control or alarm circuits are vulnerable to damage from lightning or other voltage surges, which have the more serious consequence of interrupting a vital service. Continuing efforts are being made toward improving the methods of protecting such circuits. For example, the Telephone Association of Canada is conducting a research program with the aim of measuring the characteristics of lightning transients for guidance in providing the most effective kind of protection. A thundercloud can induce a potential of several thousand volts on an exposed telephone line. A direct lightning strike to a telephone line will induce a traveling wave that will propagate along the line in both directions and can cause damage out to several miles from the source. High static voltages can be produced on exposed telephone or communications lines by dust storms and by rapid changes in the elevation of a line passing over mountainous terrain.

The above mentioned investigation[1] of the effects of lightning on telephone circuits performed during 1968 and 1969 has produced some valuable data, including the distribution of occurrence of peak surge voltage, of surge rise and decay times, and of their thermal energy content. Three types of line were examined, viz.; two open wire, four paired cable (two aerial), and four coaxial cable (one aerial). Typical surge waveforms obtained are illustrated in Fig. 7.5. There are significant waveform and magnitude variations between the three types of line, and with different soil resistivities. From the results, the investigators chose test wave shapes which simulate 99.8% of the lightning surges in open wire and cable. These are:

Facility	*Peak Voltage Amplitude*	*Rise Time*	*Decay Time*
Cable	1000 V	10 μs	1000 μs
Open wire	2000 V	4 μs	200 μs

Their conclusion was that any diodes or linear voltage devices which would withstand these test waves, would meet the stresses imposed by lightning.

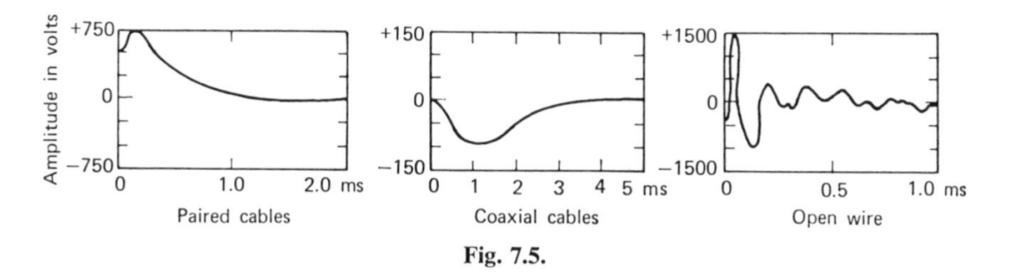

Fig. 7.5.

High static voltages can also be developed on exposed antennas and their feed lines by passing thunderclouds, and it is sometimes possible to see and hear corona discharges issuing from the top of a tower. It has been estimated[2] that man-made structures and trees transfer much more energy by corona discharge during a thunderstorm than do the lightning discharges during the same storm. The corona currents from such objects vary in value with the intensity of the storm, but lie in the range of 10 to 50 μA.

In order to divert thunderstorm energy to a ground system so as to protect communications equipment, it is necessary to have a spark gap or some other type of arrester. The term "arrester" is in fact a misnomer because the protective device actually allows the lightning current to pass through it to ground. Accordingly arresters, particularly in the low-voltage ranges, are sometimes called *protectors* or simply *spark gaps*.

In telephony a common protective device for a long time has been the carbon-block gap. The conventional 3-mil gap has a sparkover-voltage rating of 550 V. However, this threshold is too high for transistorized equipment. Furthermore, this firing voltage varies with repeated operations and during the lifetime of the gap would fluctuate between 400 and 1000 V. Steep-fronted transient voltages on a carbon-block gap raise its firing voltage to several times its nominal value. This instability, combined with relatively high impedance and voltage drop when conducting, makes the carbon-block gap unsatisfactory for protecting transistorized equipment. There is a secondary effect of repeated operation in that carbon dust deposited in the gap over a period of time can introduce electrical noise into the related circuit.

7.9 CHARACTERISTICS OF LOW-VOLTAGE ARRESTERS

A general distinction is made between arresters for the protection of low-voltage circuits like those used in telephony, communications, and control, and higher voltage types used for primary and secondary power distribution. The term "low voltage" applies to the low sparkover voltage that is necessary to the protection of circuits normally operating at low voltage and current levels.

In present-day telephone and communications practice the type of arrester—or surge protector or spark gap, as it is variously called—that is coming into common use is the gas-filled gap. It has the advantage of a low striking voltage and a low contact, or arc-drop, voltage. The latter characteristic minimizes the voltage applied to the protected equipment when the arrester is discharging a high current. A basic form of the gas-filled gap consists of two spaced electrodes sealed in a glass or ceramic capsule containing a gas. The threshold voltage, also called striking or sparkover voltage, is determined by the electrode spacing, the kind of gas mixture, and the gas pressure.

For low-voltage gaps the gas pressure is usually below atmospheric pressure. Refinements to improve performance include (a) the use of radioisotopes in the gas mixture (e.g., krypton-85) to hasten the formation of a conductive path so that it occurs in 10 to 50 nsec and (b) vaporizing metal electrodes to increase the current discharge rating to extinguish the arc. There are also gaps formed of metallized paper similar to self-healing capacitors; these have low threshold voltages and are considered fail-safe.

The striking, or firing, characteristics of a carbon-block gap and a gas-filled gap are compared in Fig. 7.6. For a given threshold voltage the gas-filled gap has a wider space between electrodes than the carbon gap, and this affords greater precision and less variation in the firing voltage. In a relatively small size a gas-filled gap can have a high rating. For example, a small model using two electrodes of 16 gauge (0.13-cm diameter) tungsten metal

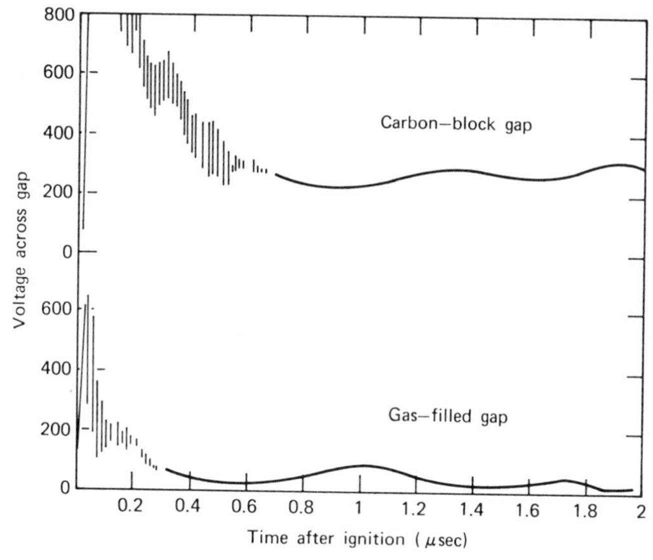

Fig. 7.6. *Comparison of the carbon-block arrester with the gas-filled gap arrester shows resistance across the gas-filled gap decreases faster and to a lower level than resistance across the carbon block. After Vodicka.*[3]

will carry a current of 80 kA for a few microseconds, after which the wire ends will vaporize, thus widening the gap so as to extinguish the arc. In order to preserve this gap for repeated operations under sustained current, a secondary device, such as a current-limiting resistor or circuit interrupter, would be used in series. Another model of a compact enclosed gap will stand a voltage surge of 50 kV and will discharge 50 kJ of energy at a current rating of 2000 A.

Gas-filled gaps are now available for a wide range of sparkover voltage, current, and energy ratings to protect various kinds of circuits and apparatus. The voltage-current characteristic of one type is shown in Fig. 7.7. The "glow," or ionization, voltage is that at which ionization of the gas begins, permitting a high current discharge. As the discharge passes through the gap, the voltage falls to the arc-drop voltage, the minimum value. After the passage of the discharge, as the current reaches zero, the sparkover voltage will rise to a value called the extinguishing, or holdover, voltage, where it would strike again if the overvoltage recurred.

The manner in which the sparkover voltage varies with the steepness of the applied wavefront for several models of gas-filled arresters with different nominal sparkover ratings is illustrated in Fig. 7.8. This type of arrester is used in communications systems as well as control, signaling, and alarm circuits. The characteristics of these devices vary somewhat in design and

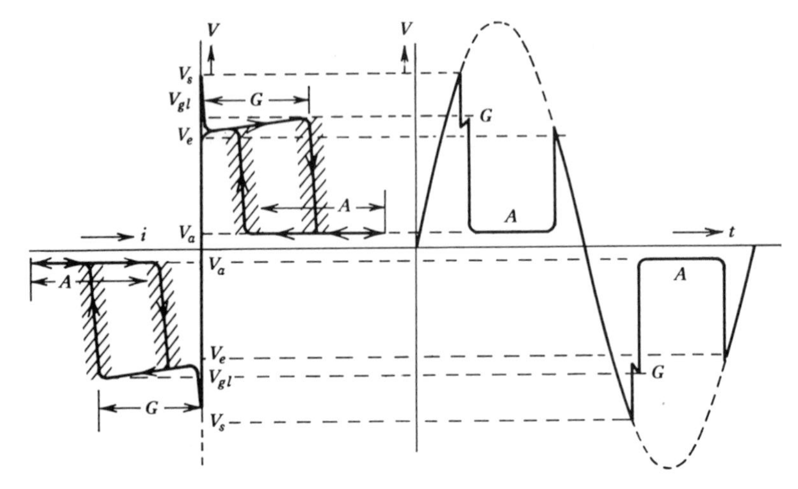

Fig. 7.7. *Voltage-current characteristic and voltage variation across the protector. Key: V_s, striking voltage; V_{gl}, glow voltage; V_a, arc voltage; V_e, extinguishing voltage; G, glow range; A, arc range. Courtesy of Siemens, West Germany.*

rating from manufacturer to manufacturer and to suit various applications. The performance of a gap varies with the kind of circuit loading it is protecting and the normal current type (ac or dc). When connected between an inductive circuit and ground, for example, the arc across the gap will extinguish at a higher value of gap voltage than that for a capacitive circuit. Tables 7.2 and 7.3 list the characteristics of a range of arresters made in the United States and Germany, respectively.

TABLE 7.2 CHARACTERISTICS OF SURGE PROTECTORS MANUFACTURED IN THE UNITED STATES[a]

MSP Model No.	Nominal dc Sparkover Voltage ±15%	Impulse Sparkover Voltage[b] at 10 kV/μsec Rate of Rise	Impulse Sparkover Voltage[b] at 500 V/μsec Rate of Rise	Minimum Holdover Voltage	Wire Leads
2001-01	230	730	540	115	None
2001-02	350	870	670	175	None
2001-03	470	1030	810	235	None
2001-04	800	1390	1140	400	None
2001-06	230	730	540	115	Yes
2001-07	350	870	670	175	Yes
2001-08	470	1030	810	235	Yes
2001-09	800	1390	1140	400	Yes

[a] MSP gaps, all $\frac{5}{16}$ in. in diameter, $\frac{11}{32}$ in. long. Data supplied by Joslyn Electronic Systems.
[b] Taken from the volt–time curves shown in Fig. 7.8.

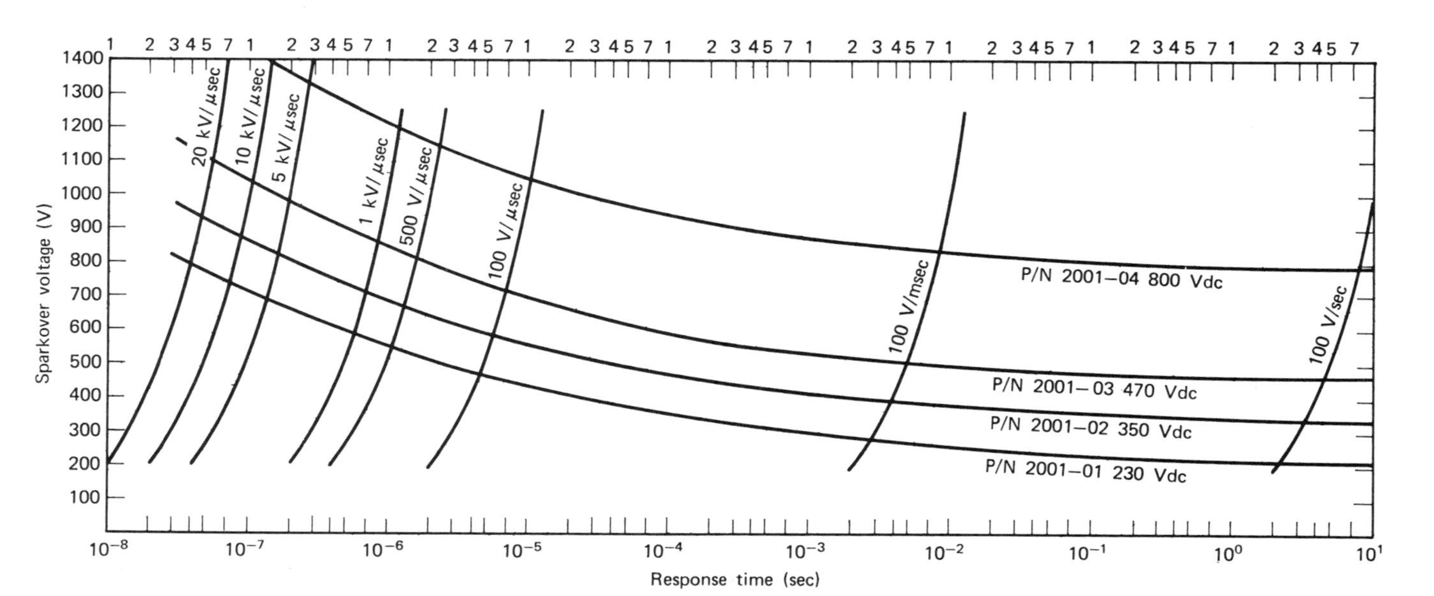

Fig. 7.8. *Response characteristics of several small gas-filled gaps for various wavefront slopes. Courtesy of Joslyn Electronic Systems.*

TABLE 7.3 CHARACTERISTICS OF SURGE PROTECTORS MANUFACTURED IN WEST GERMANY[a]

Parameter	Type: A2–B470,[b] B2–B470[c]	Type: A2–B800,[b] B2–B800[c]
Rated dc striking voltage V_{rs} (V)	470	800
Voltage tolerance (for 95% of measured values) (%)	±15	±15
For use with mains voltage (Vac)	220	380
Surge striking voltage V_{ss} (kV)	<3	<4
Maximum follow-on current for 1 sine half-wave at 50 Hz (A)	35[d]	35[d]
Rated ac discharge current (A)	5	5
Insulation resistance at 100 Vdc (Ω)	$\geq 10^{10}$	$\geq 10^{10}$
Capacitance (pF)	<2	<2
Weight (g)	1	1

[a] Manufacturer: Siemens, West Berlin.
[b] Without leads.
[c] With leads.
[d] An appropriate series resistance must be used to ensure that this value is not exceeded (*VDR*-resistor).

There are several ways in which components are associated with gas-filled gaps to make them more effective for specific applications. Some of these are the following:

1. A gap with current-limiting resistance in series.
2. A gap with both resistance and inductance in series to limit the current.
3. A dc model comprising a silicon controlled rectifier, supplemented by a shunt spark gap to prevent over load of the rectifier, and a series relay coil to interrupt the protected circuit in case of a sustained follow current.
4. Three-element gaps in a single capsule for protecting both sides of a balanced circuit to ground.
5. A gap with a fuse in series.

7.10 APPLICATIONS OF LOW-VOLTAGE ARRESTERS

As the striking voltage of arresters is dependent on the wave shape of the overvoltage, it would be desirable to know this shape when selecting an arrester. It would also be desirable to know the amount of follow current the arrester would have to carry and the number and rate of repeated operations.

Such information is generally not known with any precision and accordingly must be estimated. Also the wave shapes of voltage surges due to lightning or other causes vary with each occurrence.

For balanced circuits, such as telephone lines, drainage coils may be installed in series with the arrester between each line and ground, as illustrated in Fig. 7.9. This tends to make both arresters strike at the same time

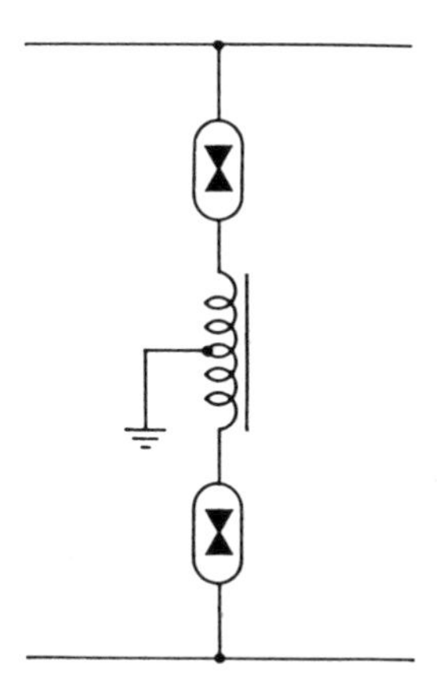

Fig. 7.9. *Drainage coil in series with gaps.*

and to retain the circuit in a balanced and operating condition even though the protective gap has operated.

An example of arresters in a video-telephone cable is shown in Fig. 7.10. This cable has active equalization circuits and carries dc power to energize them. The lightning protection is provided by a total of 20 varistor diodes at the input and another 20 at the output, this quantity being necessary to minimize the shunting effect on the bias and signal voltages.

An example of the protection of a buried communication cable is shown in Fig. 7.11. Overhead lines being transposed to underground cable are shown in Fig. 7.12, where the arresters are terminated on the metallic cable sheath. An alternative would be to terminate the arresters on a grounding rod or other buried ground conductor, provided the cable sheath was connected to the same system.

In the general case of a buried cable with metallic sheath, a fault current flowing in the sheath due to lightning or other causes will produce a magnetic flux encircling the cable, and this in turn will induce a voltage on the cable conductors within the sheath. Were it not for the voltage drop along the sheath, the voltage of sheath and inner conductors would be the same. As it is, there is a voltage difference between sheath and conductors equal to the

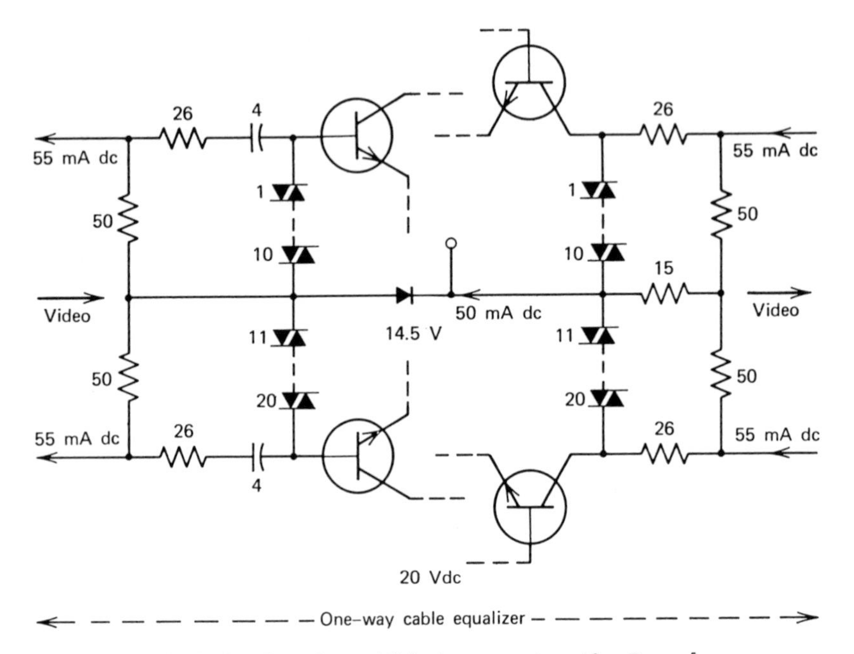

Fig. 7.10. *Powering and lightning protection. After Brown.*[5]

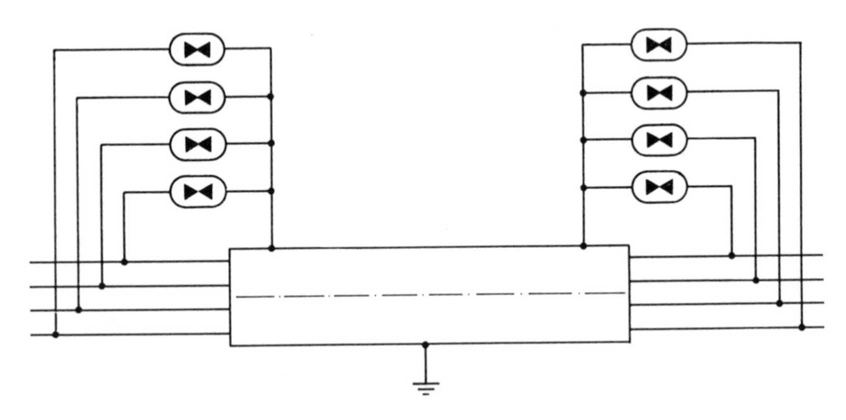

Fig. 7.11. *Protection of buried cable against induced surge voltages.*

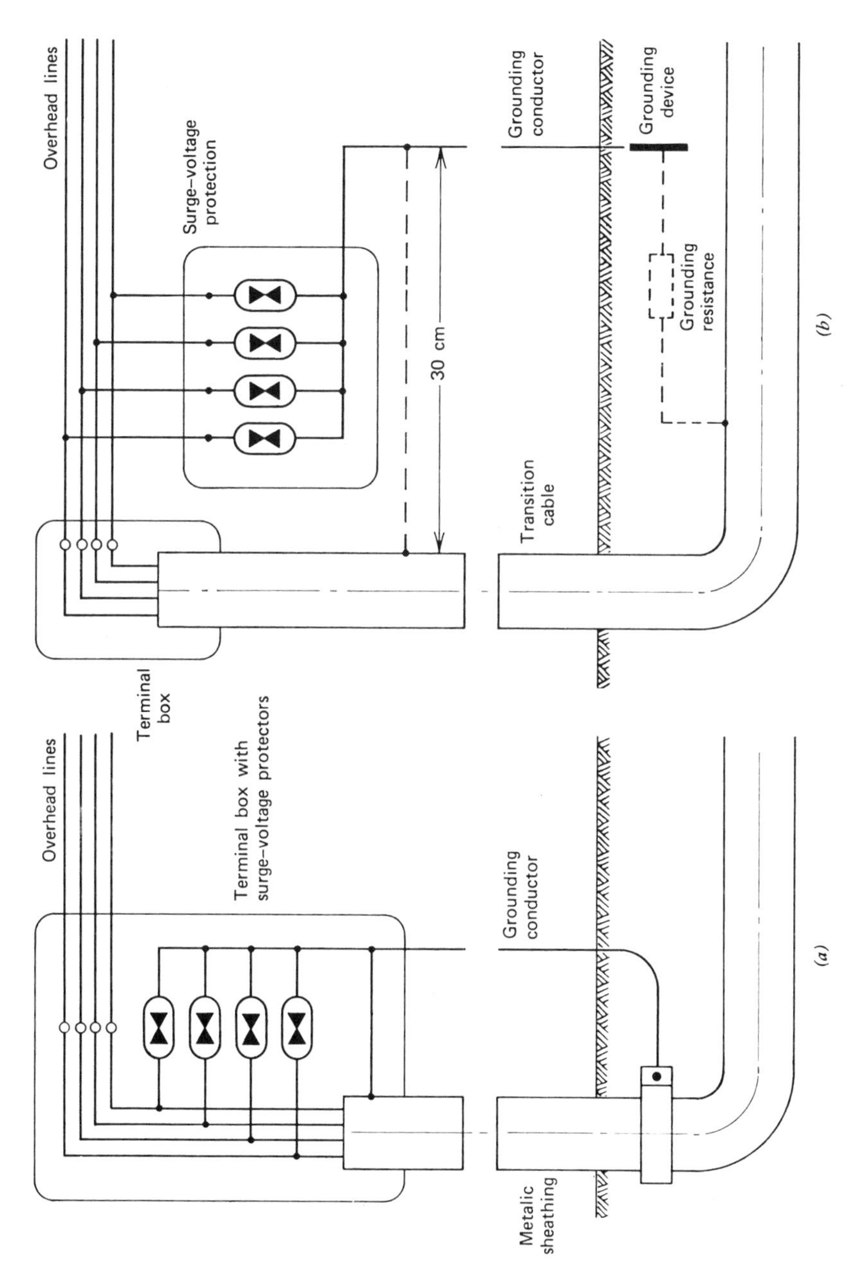

Fig. 7.12. *Connection of surge-voltage protectors at the transition from overhead line to buried cable: (a) correct: short leads, protectors connected directly to the cable sheathing, minimal grounding resistance; (b) incorrect: long leads to the protectors, often multiple grounding points, high grounding resistance of the separate grounding device. Courtesy of Siemens, West Germany.*

voltage drop along the sheath. The higher the general earth resistance surrounding a buried cable, the greater the proportion of fault current that the sheath would carry due to a nearby lightning stroke; that is, the voltage drop along the sheath would be greater, and the probability of breakdown between sheath and conductors would be greater. The voltage between sheath and conductors is nearly proportional to the square root of the earth resistivity[6] and to the direct current sheath resistance. Hence in high-resistivity soil attention must be directed to the quality of cable insulation and the size of the sheath. In some cases additional insulation over the sheath and an outer copper shield are used. The other kind of safeguard is the use of arresters between sheath and conductors, so that a final design would be a compromise between insulation, sheath size, and spark-gap protection.

Where the cable sheath is insulated over considerable lengths, a ground fault current entering the sheath at one point, say at an electric power station, would be lower if a remote ground connection on the cable had high resistance. However, this would permit intermediate parts of the cable to assume high potentials with respect to ground; this is dangerous to personnel and damaging to connected equipment. Accordingly multiple grounding of the sheath at intermediate points along the cable is usually adopted, supplemented by protective gaps between sheath and conductors.

Where considerable follow current is expected through a spark-gap protector for a cable (or other circuit), the spark-gap discharge can be used to energize a relay that would temporarily short-circuit the protected circuit and thus protect the gap from failure. This is illustrated in Fig. 7.13.

Another application of protective gaps is on aircraft antennas, where they function to equalize the lightning potential between the antenna and the metal skin of the aircraft, so as to prevent the development of potential gradients between the radio apparatus and the air frame.

It is possible by design to include inherent lightning protection in some types of antennas. An example of this is a German antenna (Figs. 7.14 and 7.15) designed for VHF frequencies. The antenna consists of circular plate supported on two vertical metal rods that provide a direct path to the grounded base plate for lightning discharges. Most TV and FM broadcast transmitting antennas are inherently designed with a direct current path to the supporting mast, which is normally grounded.

An alternative type of low-voltage arrester utilizes the varistor. The varistor draws only a nominal current when subjected to its rated voltage, but its resistance decreases exponentially with an increase in voltage, thus providing a low-resistance shunt for overvoltages. Varistor-type surge protectors for low-voltage power supplies may have shunt capacitors to bypass microsecond pulses. They can pass a substantial amount of energy for their size. For example, one model for 110- to 220-V circuits can carry

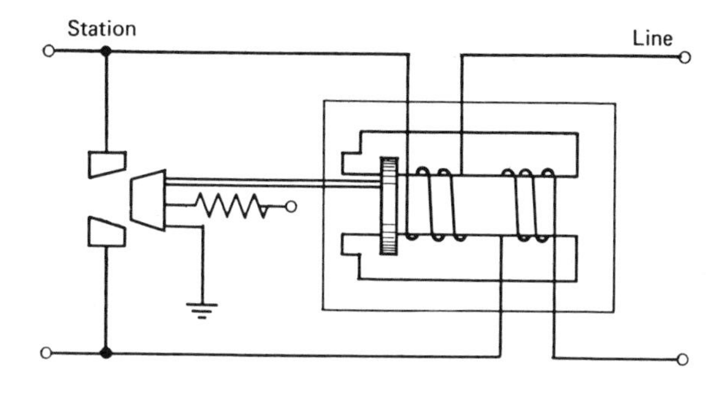

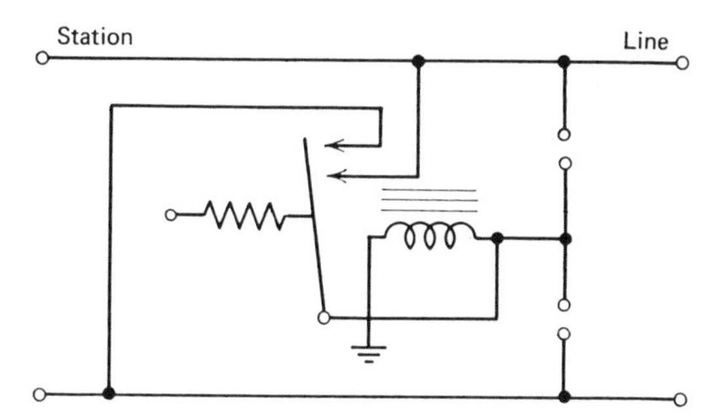

Fig. 7.13. *Principles of design of relays used for temporary short-circuiting of communications circuits during faults.*

about 90 kW over the period of a second. Models with that rating are normally mounted in a metal cabinet, whose longest dimension would be 12 to 15 in.

A type of arrester designed for insertion in coaxial cables uses magnetic force generated by the discharge to extinguish the arc (Figs. 7.16 and 7.17). An overvoltage will arc between the edge of a central spiral electrode to a cylindrical outer electrode. The surrounding coil is then energized, and its magnetic force rotates the arc along the tapered spiral electrode, lengthening it until it extinguishes itself. In this application there would be no power follow current, and the arrester would keep restriking as long as the overvoltage persisted. The arrester also contains a gas that ionizes under successive sparkovers, thereby reducing the breakdown voltage to about 60% of its initial value and facilitating the discharge of repeated voltage surges. After a 30-min period of inactivity, it will restore itself to the initial breakdown-voltage threshold.

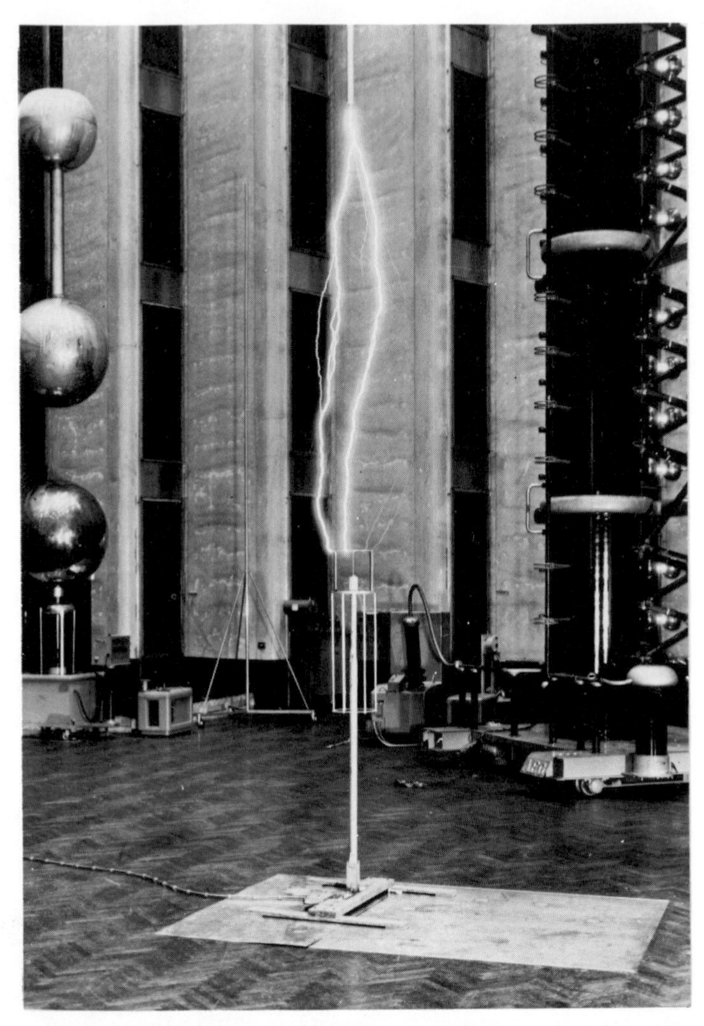

Fig. 7.14. *Lightning proof "active" receiving antenna undergoing a test from a generated discharge. The circular top plate is the antenna capacitive loading of a double folded monopole whose ground plane is the lower circular plate. In the small cylinder above the ground plate is a transistorized amplifier connected to the top plate through a series resonant circuit with a low series capacitance (high reactance). The folded monopole is resonant at the desired VHF frequency but provides a direct metallic path to ground for lightning, bypassing the transistorized amplifier. The antenna is a Rohde and Schwartz development. Photograph from Ref. 7.*

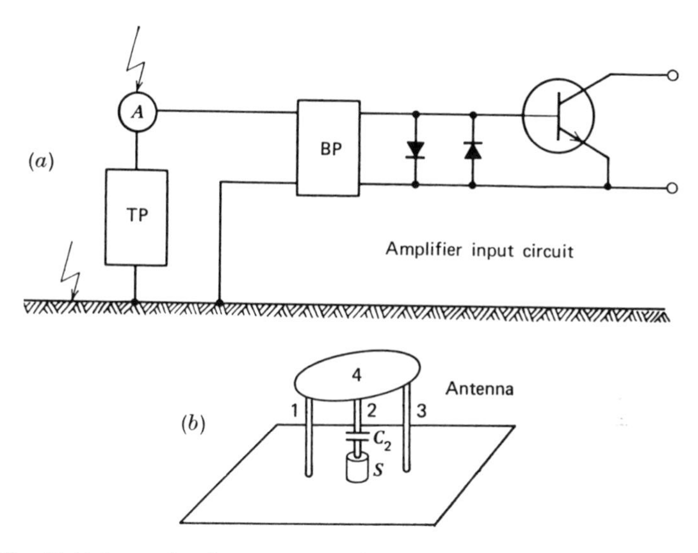

Fig. 7.15. *Lightningproof active transistorized antenna:* (*a*) *circuit outline;* (*b*) *simplified construction. Key: BP, bandpass; TP, lightning bypass; S, amplifier. From Ref. 7.*

Fig. 7.16. *Coaxial lightning arrester* (*Hy-Gain Corp.*).

Fig. 7.17. *Cutaway of coaxial lightning arrester, showing how arc is initiated between top of spiral electrode (1) and cylindrical electrode (2). Coil (3) is then energized, Magnetic force of coil rotates spark down tapered spiral electrode, lengthening it to breaking point (4) before excess current is drawn. Courtesy of the Hy-Gain Corp.*

7.11 PROTECTION OF COMMUNICATIONS CIRCUITS ENTERING POWER STATIONS

Communications circuits terminating at electric generating stations or substations pose a particular problem. The high and isolated power-transmission lines and towers are vulnerable to lightning strikes; hence if the lightning voltage surge is transferred to communications circuits, there is the risk that power follow current will flow through them also. A lightning strike to power lines which may be discharged to the power-station ground through high-capacity arresters will raise the station-ground potential to a high value for several microseconds, or longer if any power follow current flows. Any communications circuits terminating at the station, normally at near-zero earth potential, would now have a high-voltage plane below them and a high potential on any grounded parts of the equipment. There is therefore a risk of backflashover in the communications equipment, which could be followed by power current.

The use of drainage coils in balanced telephone circuits helps to inhibit power follow current and to maintain both sides of the circuit at the same potential, thus reducing the danger of electric shock to personnel. For a high degree of protection for a communications circuit entering a power station an insulating transformer is sometimes used, as illustrated in Fig. 7.18. If

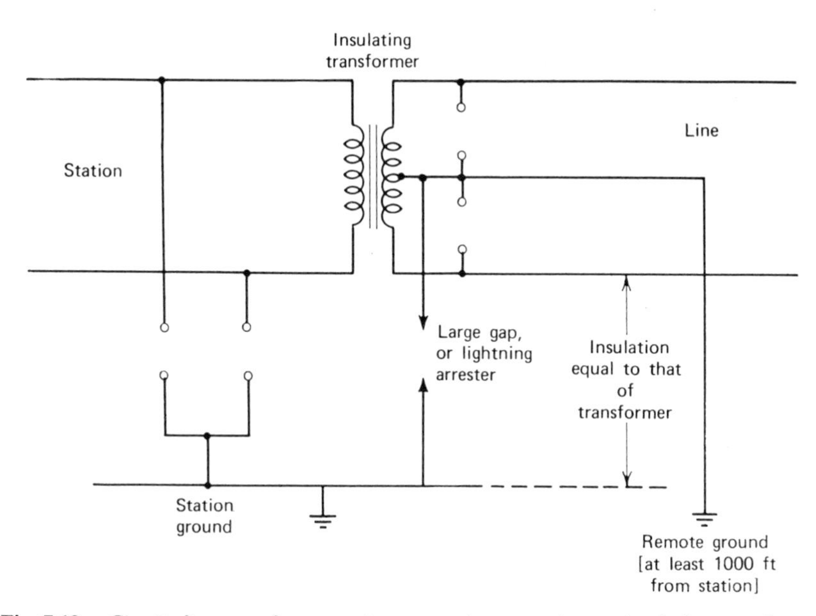

Fig. 7.18. *Circuit diagram of a protective system incorporating an insulating transformer. After Elek.*[8]

the station end of the electrical system is subjected, by lightning or other causes, to a voltage surge that discharges through arresters to the station ground, the ground system will momentarily assume a high potential. The insulating transformer will isolate the external section of the communications circuit that would maintain its near-zero earth potential by virtue of a remote earth connection about 1000 ft away from the power station. These external communications circuits are protected by low-voltage gaps (see Section 7.9) connected to the remote ground. On the station side the circuits are protected by gaps connected to the station ground. In addition, there are protective gaps between the remote ground and station ground to prevent their potential difference from becoming too great. If the external lines are subjected to lightning or other overvoltage that discharges to the remote ground, the insulating transformer prevents the temporary potential rise of the remote ground from reaching the station-side equipment.

A more sophisticated way of protecting a communications circuit entering a power station is achieved by means of a neutralizing transformer. The primary of the transformer is connected between the station ground and remote ground as shown in Fig. 7.19. Any potential difference arising between

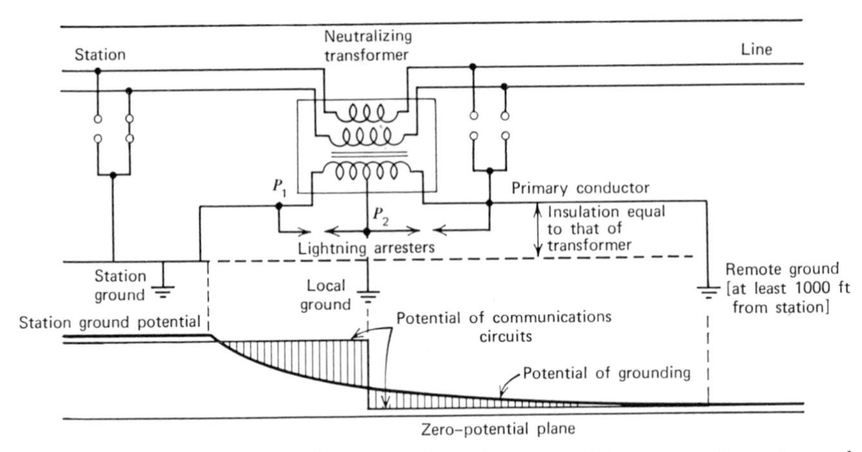

Fig. 7.19. *Diagram of a neutralizing transformer incorporated into a protective system and of the effects of the system on communication-circuit and ground potentials. Points P_1 and P_2 can be connected and the lightning arrester on the station side eliminated if the neutralizing transformer is close to the station.*

these two ground systems therefore appears across the primary of the neutralizing transformer. The two secondary windings, which have a 1:1 ratio with the primary, are put in series with the two communications lines, which are therefore raised to equal potential when a potential difference between the two ground systems is caused by an overvoltage discharge to one of them. This arrangement permits the maintenance of the communications circuit in operation under fault conditions, except when protector gaps on each side of the transformer, between the circuits and their respective grounds, might be sparking over under lightning conditions. As the potential across the transformer windings under fault conditions can reach high values, adequate insulation and/or spacing is required between entering and leaving conductors. An additional advantage of the neutralizing transformer is that it permits the transmission of direct current and the use of phantom communications circuits.

Precautions must be taken to ensure that all power-station metal, such as the enclosing fence, is bonded to the station ground so that all exposed metal assumes the same potential. Otherwise a flash could occur between a communications circuit, whose ground-system potential was momentarily raised by a gap discharge, and some nearby metal object that remained momentarily at near-zero earth potential.

7.12 RADIATION FROM LIGHTNING DISCHARGES

The total energy released in a cloud-to-ground lightning stroke can reach 250,000 J (W-sec).

A small portion of this is emitted as radiation. The average value of the ratio of radiated energy to the total energy in lightning discharges was found by Conner[9] to be about 0.007. The values for seven strokes varied between 0.011 and 0.0026. Another research group[10] found the average ratio to be 0.004. Accordingly radiated energy can reach values of 100 to 200 J. The frequency range of emissions extends from a few kilohertz to 3000 MHz. An indication of the relative amounts of radiation at several frequencies is given by Fig. 7.20 for both a ground flash and an intercloud flash.

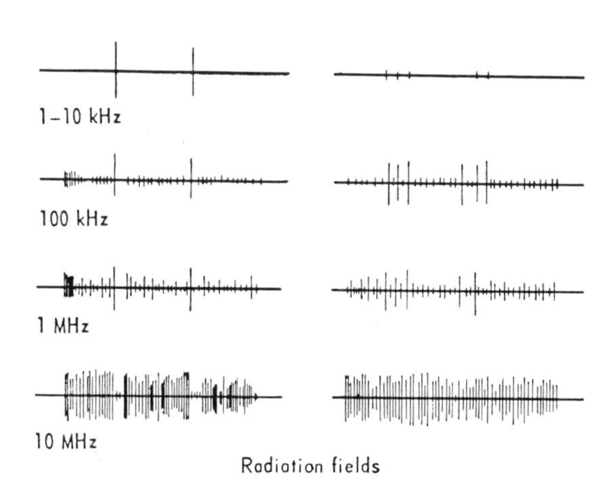

Fig. 7.20. *Radiation fields at different frequencies for typical cloud-to-ground flash (left) and typical intracloud flash (right), both at a distance of about 20 km. Amplitude scales for different frequencies are not the same. Adapted from Malan.*[11]

The radiated energy does not give rise to a need for protection, but is a source of interference to the reception of broadcasting and other radiated signals. The radiated energy also provides a means of detecting and monitoring thunderstorms at a distance, and monitoring receivers for this purpose are used by persons or companies whose activities are affected by such storms. In a case known to the author, a construction company of a very large hydroelectric generating station monitored for the approach of thunderstorms.

REFERENCES

1. E. Bennison, A. J. Ghazi, P. Ferland, "Lightning Surges in Open Wire, Co-axial and Paired Cables" paper, International Conference on Communications, Philadelphia, June 1972.
2. G. A. Isted, discussion on "Corona and Precipitation in VHF Television Reception," *Proc. IEE,* **116,** No. 10, October 1969.
3. V. W. Vodicka, "Voltage Transients Tamed by Spark-Gap Arresters," *Electronics,* April 18, 1966.
4. M. A. Uman, *Lightning,* McGraw-Hill, New York, 1969.
5. J. M. Brown, "Baseband Video Transmission on Loops and Short-Haul Trunks," *Bell System Tech. Journal,* **1,** February 1971.
6. E. D. Sunde, "Lightning Protection of Buried Toll Cables," *Bell System Tech. Journal,* **24,** April 1945.
7. G. von Flachenecker, "Eine Blitzgeschützte transistorierte Empfangsantenne," *NTZ Journal,* **10,** October 1969.
8. A. Elek, "Protection of Communications and Other Low Voltage Circuits from Station Ground Potentials During Faults," *Ontario Hydro Res. News,* **12,** No. 1, January–March 1960.
9. T. R. Conner, Los Alamos Scientific Laboratory Report LA-3754, December 5, 1967.
10. E. P. Krider et al., "Peak Power and Energy Dissipation in a Single-Stroke Lightning Flash," *J. Geophys. Res.,* **73** (1968).
11. D. J. Malan, *Physics of Lightning,* The English Universities Press, London, 1963.

CHAPTER 8

Protection Systems for Buildings

8.1 TALL STRUCTURES

Whether a large building needs a proper protective grounding system depends on many factors. One method of assessing the need has been incorporated into the British code,[1] which uses an index derived from evaluating six conditions: type of construction, its contents, its degree of isolation, the type of terrain, the building's height, and the number of thunderstorm days in the locality. In arriving at a decision to provide protection or not the index is compared with a critical value.

Usually other factors must also be considered: public safety, whether young children or elderly people frequent the structure or area, and the cost of insurance against lightning damage. If the building contains flammable or explosive material, there is no doubt that it should be protected with a well-designed system even if the incidence of lightning storms is low.

The probability of a structure's being hit by lightning depends on its height in relation to surrounding objects, the intensity of the cloud charge, and the geographical location, as described in Section 3.7.

Information was compiled in 1942 by Westinghouse[2] on the frequency of strokes to tall structures (i.e., masts, chimneys, building towers, and office buildings). The data are tabulated in Table 8.1 and plotted in Fig. 8.1. The curve indicates that the frequency of strikes increases linearly with height up to 500 ft and then begins to increase at a greater rate. This is thought to be the height at which positive leaders proceed upward from the elevated air terminal. Figure 8.1*a* shows that an air terminal of large superficial area is more vulnerable to strikes than a thin mast.

TABLE 8.1 RECORD OF THE NUMBER OF TIMES OBJECTS OF VARYING HEIGHTS ARE STRUCK[a]

Object and Location[b]	Height (ft)	Number of Years	Times Struck	Average per Year
Mast at power substation, Philadelphia	80	4	1	0.25
Ten fire towers, Western Pennsylvania	100	1	2	0.2
Radio tower, Pittsburgh	100	3	1	0.33
Radio tower, Cleveland	300	1	1	1.0
Radio tower, Cleveland	300	1	0	0
Radio tower	360	3	6	2.0
Cathedral of Learning, University of Pittsburgh	535	3	6	2.7
Smokestack, Great Falls, Montana	545	2	1	0.5
Smokestack, Anaconda, Montana	565	2	5	2.5
Empire State Building, New York	1250	3	68	23

[a] Data from Ref. 2.
[b] These objects are in regions of isoceraunic levels varying from 25 to 45 storm days per year.

The Westinghouse study compared the vulnerability of structures having 1.0% exposure (from being completely shielded) with that of structures having only 0.1% exposure. It concluded that buildings with 1.0% exposure would be struck once every 200 to 400 years; those with 0.1% exposure would have one stroke every 2000 to 4000 years. It was further concluded that for electrical systems with numerous substations it is sound economy to install air terminals that would provide sufficient shielding to reduce the exposure to 0.1%.

8.2 BUILDINGS OF MASONRY OR WOOD

Considering the air-terminal portion of a building-protection system, short vertical rods widely spaced on a large building roof are of little value because their individual attractive zones cover only a fraction of the total roof area. A more practical method is to use horizontal conductors. The corners of a flat roof and the peak of a sloping roof are the likely points of a strike because of the concentration of charge at such extremities. Accordingly a lightning conductor should be strung along the peak of a sloping roof and

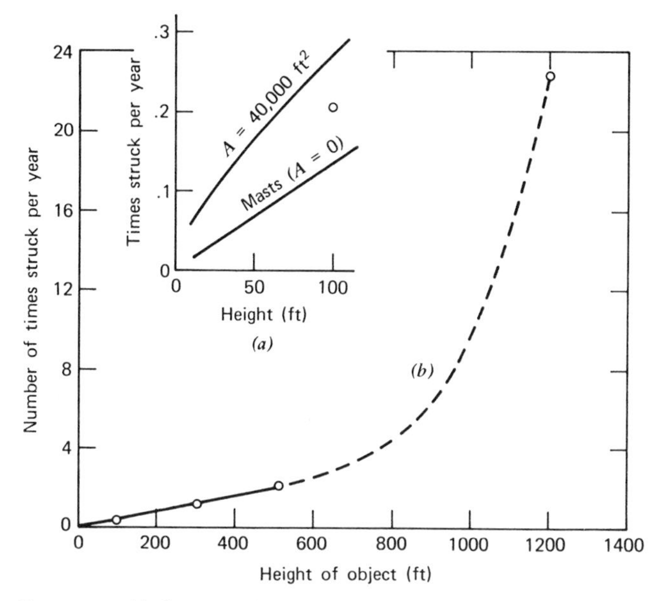

Fig. 8.1. *Frequency of lightning strike as a function of object's height: (a) calculated effect of the area of a structure at various heights; (b) curve plotted from actual records. From Ref. 2.*

around the perimeter of a flat roof. Some flat roofs have a metal railing or fence around the perimeter. This will make a good air terminal provided it is continuous or well bonded together. On large flat roofs the perimeter conductor should be supplemented with a grid of horizontal conductors spaced about 50 ft apart. Examples of systems for flat and sloping roofs are shown in Fig. 8.2.

Protrusions on flat roofs, such as chimneys or elevator penthouses, should themselves have a conductor with a connecting lead to the other roof conductors. Similarly poles, masts, and the like, if nonmetallic, should carry a conducting wire down to the main air-terminal system at roof level. If they are metallic, they should be bonded to the common air terminal or to a down conductor.

A typical grounding system for a masonry building is shown in Fig. 8.3.

8.3 STEEL-FRAME OR REINFORCED-CONCRETE BUILDINGS

If a building has a steel frame extending to the roof line, short air-terminal rods can be bonded to the metal. Similarly air-terminal rods can be bonded

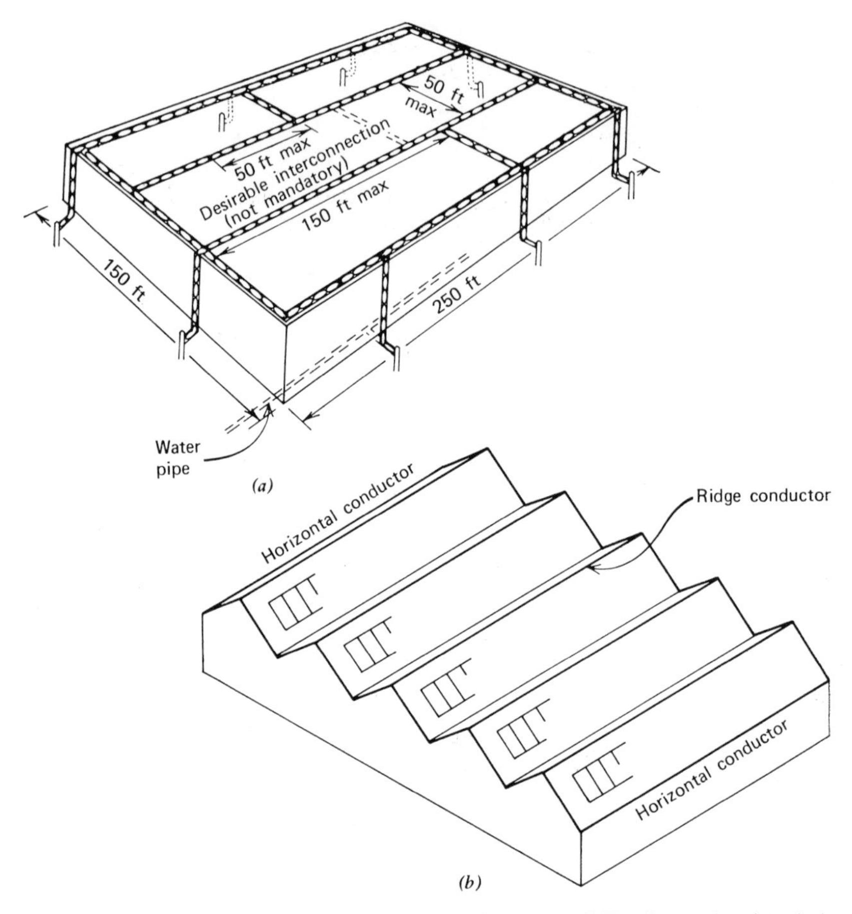

Fig. 8.2. *Lightning protection of large roof: (a) after National Fire Protection Association code[3]; (b) after British code.[1]*

to the topmost reinforcing steel in a building with reinforced-concrete walls or columns, provided the reinforcing steel is bonded together to form continuous down conductors.

Sheet-metal roofs of copper or lead with soldered joints are complete air terminals and need only be connected via proper down conductors to the earth system. Galvanized-iron-sheet roofs are likewise satisfactory if the joints are soldered. However, if overlapping sheets are joined by bolts, sparking can occur between sheets, making such a roof unsatisfactory for buildings containing explosive materials.

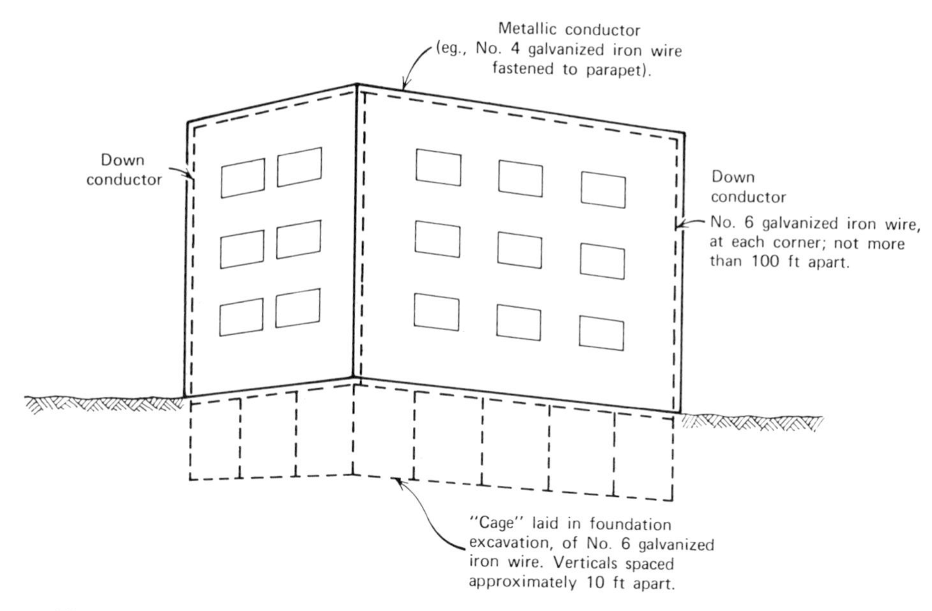

Fig. 8.3. *Practical "CAGE" grounding systems for Masonry building. An equivalent system of buried radial conductors may be used.*

If thin metal foil, 1 mm thick or less, is used in the roof covering for lightning protection, it can be pierced by the heat from the lightning current; this may or may not cause a fire, depending on the flammability of the other roofing material. This latter point will determine whether the metal foil is preferable to a regular air terminal of horizontal conductors. Asphalt or bituminous felting material is unlikely to be ignited by a lightning discharge according to tests made in England.[4]

A sketch of a grounding system for reinforced-concrete buildings is shown in Fig. 8.4.

This is a drawing of the grounding system for a large radio and television production center with a 26-floor office tower rising up from the central area. The air-terminal rods can be seen on the penthouse and building roofs, and there are groups of down conductors going down along each pylon or column. There is some justification for air terminal rods on very tall structures, to influence where the lightning will strike. The concrete-reinforcing bars are bonded to the down conductors at each floor level. The down conductors are connected to a buried electrode system consisting of six copperclad ground rods 1-in. in diameter and 10 ft long, spaced 150 ft apart. The rods are interconnected by a copper-ring bus, and the whole system is connected

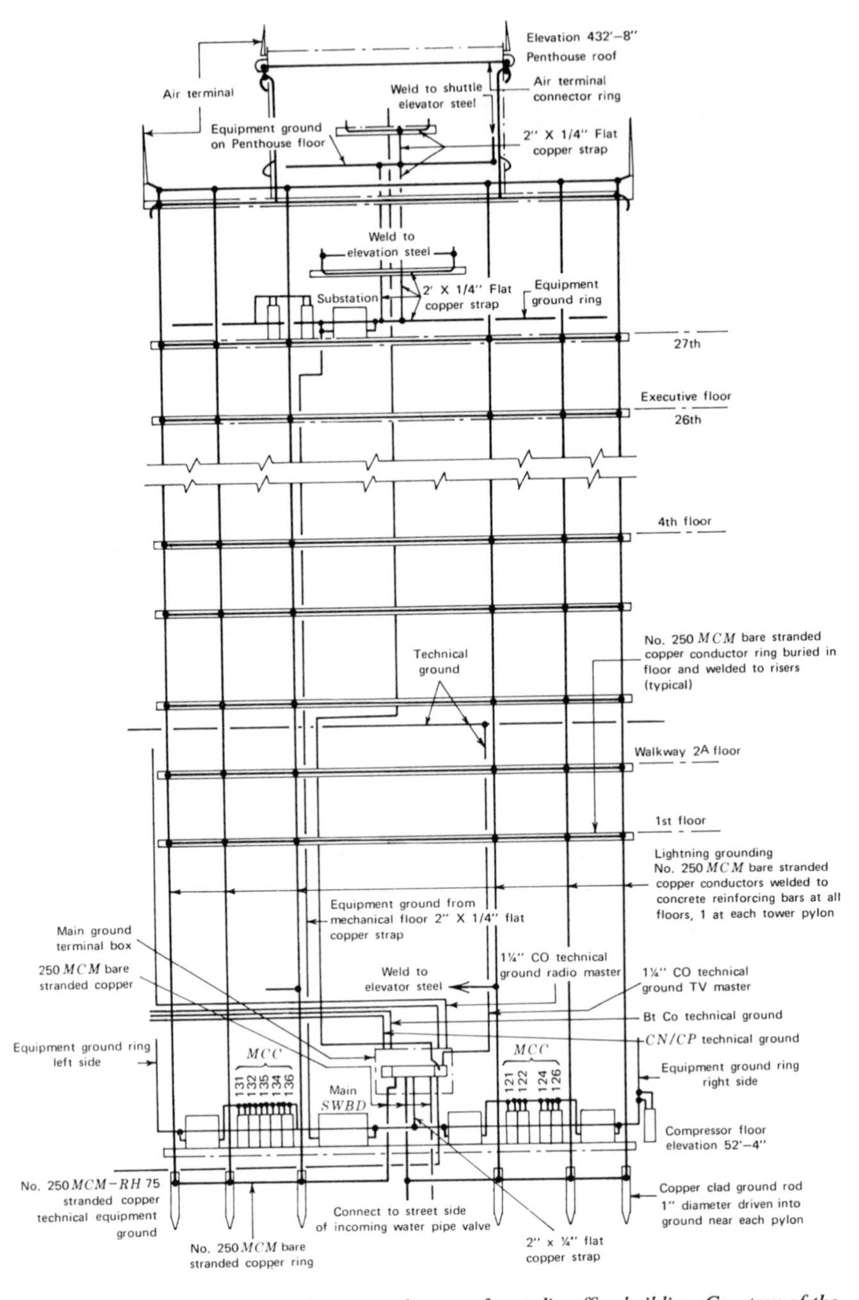

Fig. 8.4. *Grounding diagram of the central tower of a studio-office building. Courtesy of the Canadian Broadcasting Corporation.*

to a main ground terminal box at the base of the building. A connection is also made between the ground terminal box and the water main supplying the building.

All the technical equipment in the building is bonded together and brought to the main ground terminal box by separate down conductors. In a similar way separate grounding circuits from the radio system, television system, and common carrier cables are terminated in the main ground terminal box. The building foundation rests in clay soil, and the resistance of the ground system will depend on the material and effectiveness of the backfill that will bury the water pipe and connections to the ground rods. The estimated resistance of the system was 4 Ω. However, after completion of the building the ground-system resistance was measured and found to be considerably less.

The building ground terminal was bonded to the city water system by welded joints so that it was not practical to separate the two systems for measurements. The combined systems measured less than an ohm with respect to a "remote" ground; but it is difficult to have a truly remote ground in a built-up city area with many buried metal objects such as water, sewer, and gas pipes, electrical conduits, etc. It was concluded, however, that the grounding system had a very low resistance to earth and can be considered as excellent from a protection viewpoint. The soil adjacent to the building has a resistivity of approximately 100 meter-ohms.

8.4 DOWN CONDUCTORS

To conduct the lightning current from the air terminal to the earth system requires, of course, a metallic conductor from the top of the building to the bottom. If the building has no outer metallic fabric, as described in Section 8.3, then down conductors of copper wire or strap should be run from top to bottom either inside or outside the building, but not in a metallic conduit, nor in a duct of combustible material. Two such conductors are sufficient for buildings up to 100 ft square, four are sufficient for buildings of 200 ft to a side, and so on. The number is somewhat arbitrary, but it is advisable to keep the horizontal run from the point of strike to the down conductor to less than 100 ft.

The size of the down conductors may be No. 6 AWG wire or strap 1 in. wide by $\frac{1}{16}$ in. thick. However, for mechanical strength to permit secure fastening—and to permit soldering, brazing, or bolting of bonding leads—conductors of double that size are more practical.

In countries having a national electrical code the size of conductors and of earth electrodes, and the type of connections will be specified. This volume follows the principles of major electrical codes, but for specific conductor sizes and installation details the applicable electrical code should be consulted.

8.5 INTERNAL BONDING

Having a building-protection system obliges one to either bond all metallic elements like pipes and machinery to it or isolate these items so that no side flash to them will occur. Side flashes from a conductor carrying a lightning current to some other metallic element in the building are dangerous to persons and can cause serious property damage, explosions, or fire. The potential to earth from the top of a tall building, even with down conductors to a grounding system, can reach a high value due to the impedance of the protective system. A formula has been derived[4] to estimate the clearance D, in meters, required to isolate building internal metal parts from a side flash:

$$D = 0.3R + \frac{H}{15}n$$

where R = combined grounding resistance, in ohms,

H = height of structure, in meters.

n = number of down conductors connected to a common air terminal.

For buildings that are effectively metalclad for lightning protection purposes, including steel-frame buildings, the factor n can be given a high value, so that $D \simeq 0.3R$ meters. The value of the breakdown voltage for air used in this formula is 5 kV/cm for spacings greater than 50 cm.

Effective isolation of internal metal in large buildings is considered rather impractical because it puts too great a restriction on the placement of pipes, ducts, and the like. It can be applied in dwellings where the location of metallic elements can be more easily controlled. In dwellings a clearance distance of 10 ft is satisfactory, assuming that the resistance of the grounding system is about 10 Ω.

In a large building with a protective grounding system the bonding of internal metal is the preferred alternative to the more difficult isolation method. If the floors of the building contain reinforcing bars or other metal, such metal should be bonded at joints so as to reduce the potential drop across the area. Side flashes within a concrete floor can fracture the concrete.

If metal piping, ducts, and the like within a reinforced concrete or metal-frame building do not continue to the earth, or are not connected to grounded objects, they would be shielded by the outer metal fabric of the building and the potential between the outer "cage" and the internal metal would not be sufficient to cause side flashes. However, since this condition of isolation from earth rarely occurs, bonding is nearly always required. Conduits, cable sheaths, pipes, or ducts that run vertically through the building should be bonded to the building's outer metal or down conductor at the top of the building, at the bottom to the buried earth systems, and at intervals of five storeys in between. This vertical interval is arbitrary, but it is certain that bonding on every floor is too frequent and that intervals of 100 ft are probably too great.

8.6 THE EARTH TERMINAL

Having a suitable air terminal and a continuous current path to earth via down conductors or the outer metal fabric of the building (to which all internal metal is bonded), the next consideration is the buried earth-electrode system. The types of buried earth-electrode systems, employing rods, horizontal conductors, or both, are described in Chapter 4. The particular situation at each building will determine the most suitable arrangement. If space surrounding the building is limited, long rods driven into virgin soil are the recommended form of ground system. They would be connected by buried conductors to down conductors or the outer metal fabric of the building, using brazed, welded, or bolted joints. If space permits, earth-resistivity measurements should be made first and then a system of rods installed in accordance with the principles outlined in Section 4.10. The objective should be an earth-system resistance of 10 Ω or less. The resistance of the system to earth should be measured as described in Section 4.5, if space permits, before connecting to the building metal, or down conductors.

If there is insufficient area even for ground rods, the "cage" method described in Section 7.3 should be employed. Conductors of No. 6 AWG wire (or of thicker wire if greater strength is desired) should run vertically from the bottom of the excavation to the surface, spaced a distance equal to the depth of the excavation. A horizontal conductor just below final grade level should join them at the top, and another horizontal conductor should interconnect them just above the bottom of the excavation, so as to form the cage. In this situation the degree of contact between earth and conductors will depend on the compactness of the backfill. Accordingly in calculating the resistance to earth some judgment will be necessary in applying the method described in Section 4.10, which assumes complete contact with the soil. Also, if space and surface conditions permit, the resistance of the system to earth should be measured after backfilling, but before connection to the building metal or down conductors.

If the excavation is in gravel or rock, the problem of obtaining a low-resistance ground system is magnified. A rather complete cage including the bottom of the excavation is recommended so as to multiply the chances of contact with the soil material, water pools, and the like, and to provide electrical capacitance to the surrounding rock mass. Conductors should be laid vertically down one side of the excavation, across the bottom, and up the opposite side. A similar row of conductors should be installed between the other pair of sides so as to form a large basket grid. The conductor spacing should be equal to the depth of the excavation so as to compensate for the high soil resistivity and presumably a poorly compacted earth and rock backfill or gravel backfill. Similarly an extra horizontal conductor should be added midway between the top and the bottom of the excavation. If any loam or clay can be mixed with the backfill, this will help to lower the resistivity.

It would be difficult to measure the resistance to earth of such a system if there were only shallow surface soil in the surrounding area. To calculate its resistance one could proceed in accordance with the method of Section 4.10 or extrapolate from the curves, but should make some arbitrary assumptions to allow for the actual conditions. Let us say the excavation is 100 ft square and 10 ft deep, and the average conductivity of the rock material is estimated to be 2000 Ω-m. Although extra conductors have been used, let us assume an effective lateral spacing of 10 ft between vertical runs so that there would be 40 verticals in all. From the example in Section 4.10 the resistance of this system would be approximately 45 Ω.

Now if nearby there were a buried water main whose resistance to earth is 30 Ω, the combined resistance of the two systems together would be about 20 Ω, which is a fairly satisfactory value for rocky soil conditions.

A well that has to be drilled for the principal or supplementary water supply can be a benefit for grounding purposes. For example, a 100-ft well with an 8-in. steel casing sunk in rock material where ρ is 2000 Ω-m will have a resistance to earth of about 65 Ω.

8.7 CONNECTION TO BURIED PIPES

Any buried metallic pipes or conduits that enter the building must be included in the bonding of all internal metal to the buried earth-electrode system. If this is not done, there is a risk that a lightning discharge potential would cause electrical breakdown of the soil between the lightning ground system and the buried pipe, with resulting damage due to arcing. On the other hand, a new problem arises if gas or petroleum products are being carried in the pipe, which might be sectionalized by insulating couplings for cathodic protection. If such a pipe is made part of the buried ground system, a high lightning potential could cause a flashover across the insulation and possibly cause an explosion. To prevent this happening the insulated pipe couplings should be bridged by encapsulated spark gaps set to flash over before the breakdown voltage across the coupling is reached.

It will be evident that if hazardous pipes like these are already installed before the building is planned, it would be rather impractical and costly to install discharge gaps across the insulated couplings. A logical alternative would be to space the buried earth electrodes for lightning sufficiently far from the pipe to prevent an earth breakdown between them. If a branch from the hazardous pipe is to enter the building, it should be isolated with a length of nonmetallic pipe. The required separation between earth electrodes and other buried metal can be calculated as described in Section 7.2.

8.8 HAZARDOUS BUILDINGS

Buildings that contain explosive or highly flammable material require special measures. As in all cases, the building must be the highest object in the immediate vicinity or outside the protected zone of neighboring structures to warrant attention at all. The objective in protecting hazardous buildings is to keep discharge currents outside the structure. This can be done by installing a shielding grid of conductors over the roof, but isolated from it and connected to down conductors that are also isolated or insulated from the structure. The number of down conductors is made double the usual number, to reduce the potential, and they should connect at the base to a buried ring conductor that bonds together all earth electrodes. The alternative is to follow conventional practice, but use more air-terminal wires and down conductors, and pay particular attention to bonding all internal metal both together and to the down conductors.

In all protected structures one must ensure that any metallic equipment, pipes, and the like added subsequently are bonded to the existing grounding system.

Where possible, hazardous buildings should be located so as to minimize the risk of lightning: in low-lying areas rather than on heights, and where there are some scattered trees rather than completely in the open. A deep layer of earth is preferable in that a low ground-system resistance can be attained. However, a high degree of protection is possible even if the soil resistivity is high.

Where it is convenient and permissible, a metalclad building including a roof bonded by welding or soldering provides a conducting shroud to carry the discharge current to earth, providing, of course, that there is an adequate buried-ground-electrode system. In deep soil the buried ground system should comprise a ring of ground rods around the building, separated by their depth at least and joined by a buried ring conductor. The metal cladding at the building base should be connected to the buried system at each rod. Although a ground-system resistance of 10 Ω or less is desirable, the primary objective is to equalize the potential around the building. If water, electricity, or other services must be taken into the building, these should be carried in metal conduits that are electrically bonded to the buried ground system.

A steel-frame or reinforced-concrete building with suitable bonding of the metal will likewise provide a conducting path to a buried ground system, provided the building has a roof with a metallic grid within it or a continuous metal roof, electrically bonded to the wall metal.

If the roof is of continuous metal and the walls are not electrically conducting, down conductors bonded to the roof metal at each corner of a small building or at 100-ft intervals on a large building should be carried down the walls, fastened at short intervals, and connected to the buried earth system.

For buildings that are not metalclad and contain explosive gases or materials, a canopy or shielding wires isolated from the structure is recommended to minimize the risk. The wires should be suspended above the roof, from masts, and down conductors at each mast should connect to the buried earth system. The wires and down conductors should be a minimum of 6 ft from the structure,[1] and the horizontal spacing between wires should be no greater than the height of the building. The size of the conducting wires, which may be of copper or galvanized iron, will be determined by the mechanical strength required, rather than current-carrying capacity (e.g., No. 6 or 4 AWG wire is suitable). A typical wire-canopy system is shown schematically in Fig. 8.5.

If the soil surrounding the structure is shallow, a number of buried radial conductors should run outward from a buried ring conductor encircling the building. The extent of the ground system necessary to attain a resistance in the 10- to 20-Ω range can be estimated from the data in Sections 4.10 and 4.12.

If the hazardous building is located in sand, very long ground rods should be used to reach well down into the moisture region. As indicated by Fig. 4.9, the resistance of a 50-ft rod is about one-third the resistance of a 10-ft rod, and in sand this ratio is likely to be greater.

If the hazardous building is situated on rock, which would be a bad choice of site, a continuous metal mesh should be laid out around the building, all down conductors being bonded to it. A cage in the basement or foundation excavation, or under the floor, as described in Section 7.3, would contribute to lowering the ground-system resistance and to protecting personnel within the building. This applies to each alternative method. However, on rocky terrain it would not be possible to reduce the ground-system resistance to 10 or 20 Ω with any reasonable area of metal mesh. Accordingly the wire

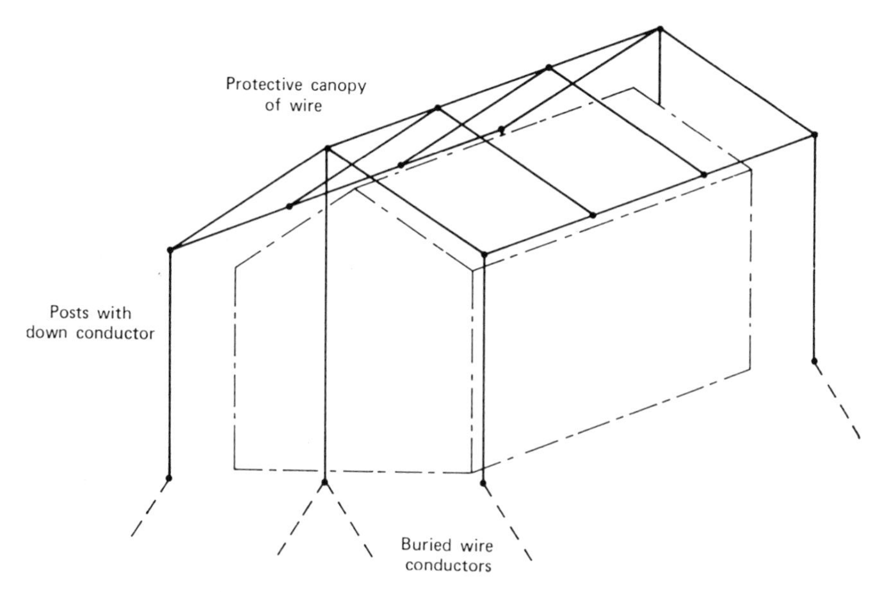

Fig. 8.5. *Protective canopy for hazardous building.*

mesh should be extended to the property limits or any reasonable area around the building, and a fence should be erected to enclose it. The fence would presumably have metal posts to which the mesh should be bonded. Appropriate signs should be attached to the fence, warning people to stay clear during thunderstorm periods. The function of the wire mesh is fourfold: to lower the ground resistance somewhat; to equalize the potential around the building; to prevent arcing due to breakdown of the rock soil under high voltage; and to protect personnel in and around the building. The size of the conductors in the mesh will be governed by mechanical strength. Galvanized-iron mesh of No. 4 or No. 6 AWG wire is suitable. The opening in the mesh may be 6 to 12 in., that is, not much longer than a person's foot.

As mentioned in Section 2.9, corona discharge from earth objects can occur under a highly charged cloud even without a lightning flash occurring. Metallic ventilating pipes or similar protrusions from the roofs of buildings can produce a corona discharge. It is possible that such discharges (normally a few microamperes) can ignite explosive vapors. Precautions that can be taken are to use nonconducting vent openings, attach corona shields to them, and put flame traps in the vent openings.

The protection of hazardous buildings should be designed only after careful assessment of the problem. A thorough study of the structure, its contents, the local terrain, and all the grounding principles described in the preceding sections is necessary. Careful measurements of earth resistivity are required as a basis for design, and the buried-earth-system resistance should be measured after installation to confirm that the expected results were obtained.

8.9 COORDINATION

It is conceivable that a tall building whose buried ground system is connected to a metallic water main supplying water to an adjacent building could constitute a danger to the adjacent building. If the terrain is rock, with a high resistivity, the potential applied to the water main from a lightning discharge at the tall building could cause injury or damage in the adjacent building if there is no internal bonding to equalize the potential throughout the structure. In such a case it would be prudent to examine the legal implications and to consider insulating the water main from the new building ground system by a nonmetallic section.

8.10 PROTECTION OF TALL CHIMNEYS AND SPIRES

Tall chimneys, if nonmetallic, are usually fitted with a metallic air terminal and down conductors. Two or three down conductors equally spaced on the wall of the chimney should be used on chimneys that are several hundred feet high so as to ensure a low reactance value. The down conductors should be bonded at 50-ft intervals by a horizontal wire of the same material. Copper wire of No. 2 AWG or copper strap would be suitable for conductors.

The design of the buried earth system should follow the principles already described for buildings, and the foundation excavation should be exploited to install a buried cage. If a building is to be constructed adjacent to the chimney, its excavation may also be used for a cage, in order to get the net resistance of the ground system down to 10 or 15 Ω. All internal metal in the building would have to be bonded together and connected to the buried ground system in order to protect the occupants as well as the building fabric. In chimneys built of reinforced concrete the bonding of the steel reinforcing would constitute a satisfactory down conductor, but this might be no cheaper than external down conductors.

The spires of cathedrals and churches and the pinnacles of other buildings are generally vulnerable to lightning strikes. They are usually protected by the three elements of a grounding system, that is an air terminal, down conductor and buried earth electrodes. If the spire is metallic then it is inherently an air terminal, to which a down conductor can be connected, brazed or welded. The down conductor may be no. 2 AWG copper, stranded for flexibility to conform to the contours of the structure, and fastened at intervals to the outer wall. The buried electrode system should follow the principles already given with the objective of achieving a resistance to earth of 10 ohms or less.

8.11 TOWER-TYPE BUILDINGS

There is a trend toward building tall towers, 1000 ft high or more, for observation decks for visitors, communications and broadcasting antennas, high-altitude restaurants, or a combination of these purposes. Such structures are, of course, vulnerable to lightning strikes. Accordingly they should be fitted with an air terminal, a continuous conducting path to earth, and a low-resistance ground system. The down conductor could be the continuous steel framing from top to bottom, reinforcing bars continuously bonded in a concrete structure, or a series of down conductors of heavy wire or strip. Either No. 2 AWG wire or $1 \times \frac{1}{16}$-in. strip can be used. Four such down conductors would be a suitable compromise between low impedance and high reliability.

The degree of protection will depend largely on the impedance of the buried electrode system. A resistance in the 10- to 25-Ω range should be the objective. The excavation for the foundation will permit a cage type of system to be installed. If the base of the excavation is not rock, a 10- or 20-ft ground rod could be driven down at each corner within the excavation and connected to the cage. If the excavation base is rock, the cage would have to be supplemented by horizontal buried conductors if the resistance of the cage alone were too high. Of course, metal water mains or a well-pipe casing may be incorporated into the ground system, if available. As was indicated for buildings in general, all metal components or fixtures in the tower structure should be bonded to the down conductors. An example of a grounding system for a large office tower building is shown in Fig. 8.4, but the relatively low earth resistivity in this case obviated the need for a "cage" arrangement.

8.12 RESIDENCE PROTECTION

Although family dwellings are not often struck by lightning, they are vulnerable when situated on large lots without tall trees or other high objects nearby. A common form of earth terminal in the home is a copper ground rod driven into the earth below the basement floor or into the earth adjacent to the foundation; it is usually supplied by the electricity supply agency. Any home owner wishing to have a suitable protective ground system should have a second driven rod at the opposite end of the house for the small dwelling in good soil; a rod at each corner if the soil is gravelly or mixed with stones, or if the dwelling has a large floor area.

If the underground water pipe entering the house is metallic, it should be included in the grounding system.

From the calculations of resistance to earth described in Section 4.8, two 6-ft ground rods in soil with a resistivity of 100 Ω-m would have a combined resistance of about 30 Ω; four such rods would have a combined resistance of 16 Ω. If we assume that there is 50 ft of water pipe at a depth of 5 ft underground and make no allowance for the water main, which might be nonmetallic, the resistance component of the water pipe in soil with a resistivity of 100 Ω-m would be 10 Ω. This value combined with, say, two ground rods would give a resultant resistance of about 8 Ω. This assumes, of course, that the electrodes act in parallel, by being bonded together by a heavy-wire conductor (e.g., No. 6 AWG) or by common connection to down leads from an air terminal. From a practical point of view, the 10-Ω resistance provided by the water pipe alone would be satisfactory.

For an air terminal, a single short lightning rod on a house roof is not very effective, because its protective zone is relatively small. A better terminal is a wire conductor along the peak of a gable roof or a grid of wires over a large flat roof. The breadth of the openings in the grid may be approximately equal to the height of the house. An illustration of a residential grounding system is shown in Fig. 8.6.

Should the dwelling be located where the surface soil layer is shallow and lying over bedrock, it would not be practical to drive ground rods. Under such conditions buried radial conductors of No. 6 AWG bare copper or galvanized-iron wire should be run out from each corner of the house to the extremities of the lot. The inner ends can be brought to a convenient common point at the floor of the excavation or under the main floor of the house if there is no excavation.

Assuming the four buried conductors average 50 ft in length (15 m) and are buried 6 in. below the surface in soil where the resistivity ρ is about 1000 Ω-m, the resistance to earth, R, of the system will be on the order of 30 Ω. In this case, if a buried metallic water pipe 5 ft below the surface and 50 ft long can be connected to the ground system, the resultant R value would be reduced to 9 Ω. This assumes that the backfill material over the pipe would also have a ρ value of 1000 Ω-m. These values are extrapolated from the curves and calculations of Chapter 4.

As many houses have a rooftop television antenna, they have an "air terminal" that is probably more of a liability than a benefit, because its

Fig. 8.6. *General grounding plan for a house. From Ref. 5.*

height makes it the likely point to be struck. The support mast of the antenna should be connected by a down lead to a ground rod or to a whole buried ground system if such exists. The live conductors of the antenna should be taken to ground by protective gaps (lightning arrester) on the antenna mast or at the nearest convenient point. If the television cable is coaxial or has a metallic sheath, this must be bonded to the grounded mast also. These measures will divert any lightning current to earth with the least risk of high potentials reaching the television receiver and any persons who might be touching it.

Protection of the home is intended more for the safety of the occupants than for the prevention of material loss. However, it has been included in this chapter because of its close relationship to the protection of buildings. The protection of temporary shelters like tents and cabins is dealt with in Section 6.7.

8.13 PROTECTION OF FARM PROPERTY

Farm buildings without tall trees nearby are relatively vulnerable to lightning strikes because of their isolation. The principles of grounding the farm home and other small buildings are the same as those discussed in the preceding section. For large barns more branch conductors are required,[5] as illustrated by the sketch in Fig. 8.7, and the principles are similar to those for wood and masonry buildings described in Section 8.2. Of course, if the building has a metal sheath, well bonded together, then one need only be concerned with the buried electrode system. It is important to bond the internal and external metallic components of the building (e.g., stanchions, conveyor tracks, eaves troughs, metal roofs) to the down conductors of the system. If there is a well with metal casing, this could be the main earth terminal; and combined with buried wires from the earth terminal of each building, it would form an effective ground system. The joints between the buried wire, well casing, and the building electrodes would have to be properly made with clamps, brazing, or soldering.

Wire fences with wooden posts or metal posts set in concrete will assume a high potential throughout their length if struck by lightning, which increases their danger to livestock in contact or near enough to receive a side flash. For protection a ground rod of 6 to 10 ft should be driven into the earth at intervals of about 150 ft beside fence posts, and all fence wires should be connected to it by a short length of conductor. Isolated trees where livestock gather during a storm can be made more safe by installing air terminals at its extremities[5] and down conductors terminating in buried wires radiating outward from the tree. The radial wires are preferable to a ground rod because they also reduce the voltage gradient along the ground.

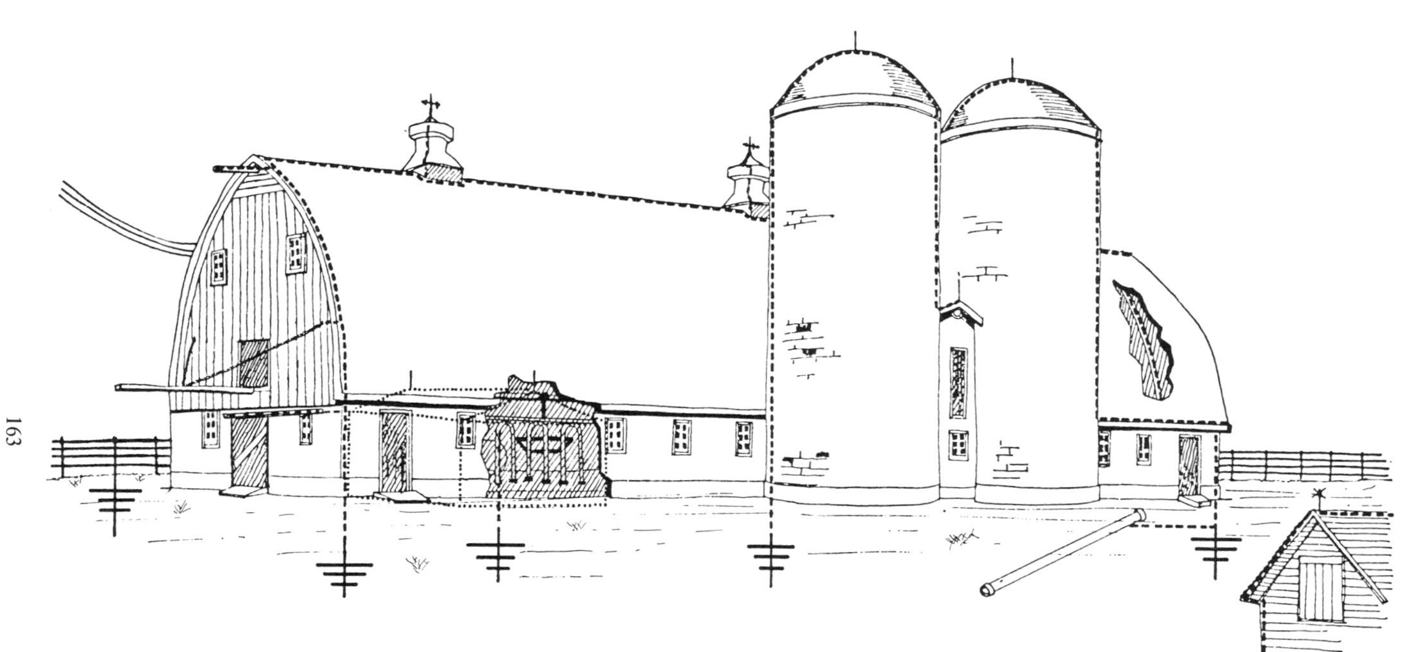

Fig. 8.7. *A general grounding scheme for farm buildings. From Ref. 5.*

8.14 HISTORICAL BUILDINGS

Custodians of valuable historical buildings are becoming more concerned about protecting them against lightning. Early in 1970 it was reported in the press that the Greek Archaeological Service began to install lightning protection at four historical sites. A tall lightning rod (presumably connected to a buried electrode system) was installed on the Acropolis to protect the Parthenon and other ancient monuments. Lightning conductors were also installed on the Temple of Aphaia and the Temple of Apollo, a medieval palace on the island of Rhodes. Other ancient sites will be provided with protection also. The protection program was initiated after a limestone column of the Temple of Aphaia, south of Athens, crumbled from a lightning stroke a year previously.

REFERENCES

1. *The Protection of Structures Against Lightning,* BSI Code of Practice, CP. 326, 1965, London.
2. Westinghouse Editorial Service, *Proc. IRE,* February 1943.
3. National Fire Protection Association, *Lightning Code 1965,* Boston.
4. R. H. Golde, "Protection of Structures Against Lightning," *Proc. IEE,* **115,** No. 10, October 1968.
5. Lightning—Protection for the Farm, Farmer's Bulletin No. 2136, U.S. Department of Commerce, Washington, D.C.

CHAPTER 9

Protection of Power-Transmission Systems

9.1 REASONS FOR PROTECTION

Since the days of Steinmetz in the 1920s many investigations have been carried out on the effects of lightning on power-transmission lines and on methods of protection. It is probably correct to say that the study of lightning for the purpose of power-line protection has been the basis of the principles that have been evolved for the protection of all other systems and structures. Maintaining the continuity of electric service and the avoidance of major damage to transformers, switchgear, and other components provide strong reasons for protecting power systems.

Although protection methods are well developed, lightning still causes frequent power interruptions. For example, in Britain about 15,000 interruptions per year occur on high-voltage lines,[1] and the average isoceraunic level there is relatively low, at 10 or less. An investigation made in Britain[1] indicated the relationship between the number of thunderstorm days and the frequency of lightning incidents (one or more circuit interruptions due to one lightning stroke) on power lines of various transmission voltages. These relationships over a 10-year period are shown in Fig. 9.1.

In the United States, where the isoceraunic level varies from nearly zero in the northwest to a maximum of 90 in Western Florida, a large amount of data on lightning strikes was collected by a committee of the American Institute of Electrical Engineers (AIEE).[2] From these data they plotted the two curves of Fig. 9.2, which indicate the number of strokes to transmission lines per year per 100 miles of line (curve B) and the percentage of these strokes that had currents exceeding a particular value (curve A). The AIEE

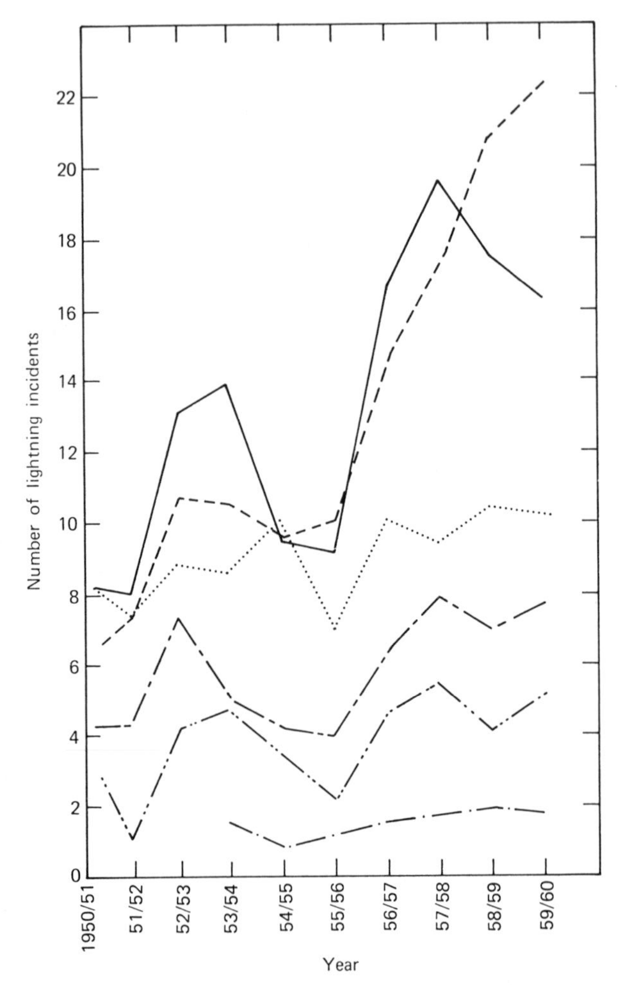

Fig. 9.1. *Annual variation in the frequency of lightning incidents and thunderstorm days. Key: -----, 11-kV system; ——, 11-kV system; —-—, 33-kV system; —··—, 66-kV system; —·—, 132-kV system; ····, annual number of thunderstorm days. From Golde.*[1]

committee ascertained from the data that in areas where the isoceraunic level is 30, about 100 strokes would occur per 100 miles of line per year.

As the transmission-line voltages have increased over the years, the lightning strikes to them have increased. This is attributed to the higher towers and wider crossarms, which reduce the shielding effect of the overhead "ground" conductor.

However, for the newest high-voltage lines, exceeding 300,000 V, the

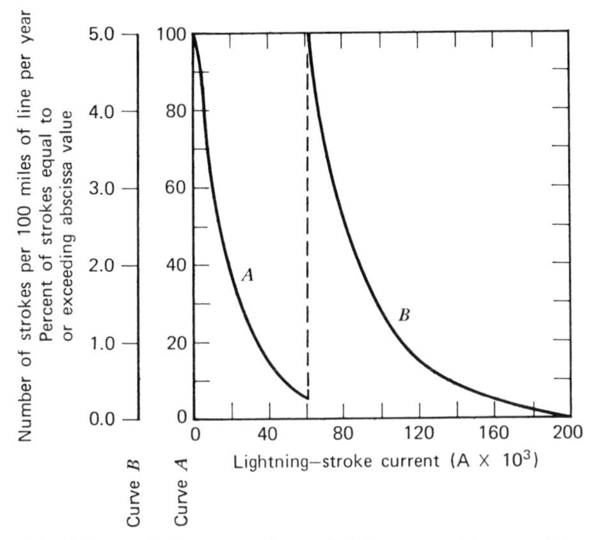

Fig. 9.2. *Lightning-stroke-probability curve. From Ref. 2.*

insulation required to withstand internal switching surges is so high that most lightning strokes can be conducted to ground without causing a flash-over or can be limited to short tripoffs if a phase conductor is struck.

The existing literature on transmission-line protection is extensive, and several of the sources are given in the reference list. This chapter will provide only an outline of the principles and methods of protection.

9.2 REVIEW OF THE DISCHARGE MECHANISM

Recent investigations[3] indicate that a typical downward stepped leader would have a potential of about 50,000 kV. Using an accepted value of 500 kV/m as the sparkover-voltage gradient for air, it follows that when the leader tip comes within 100 m of a grounded object, the main discharge would occur. The leader-tip position at this instant is called the point of discrimination[3] because prior to it the leader is independent of the impedance to earth. Immediately thereafter, as the return stroke develops, the voltage gradient of the channel increases for a period of 1 μsec or less, and one result of this is a slowing of the return-streamer velocity to between 0.1 and 0.3 the velocity of light. The velocity of the return streamer increases with increasing stroke current. The formation of the final flash (described in Section 2.5) is depicted in Fig. 9.3. The process of charge collection by both

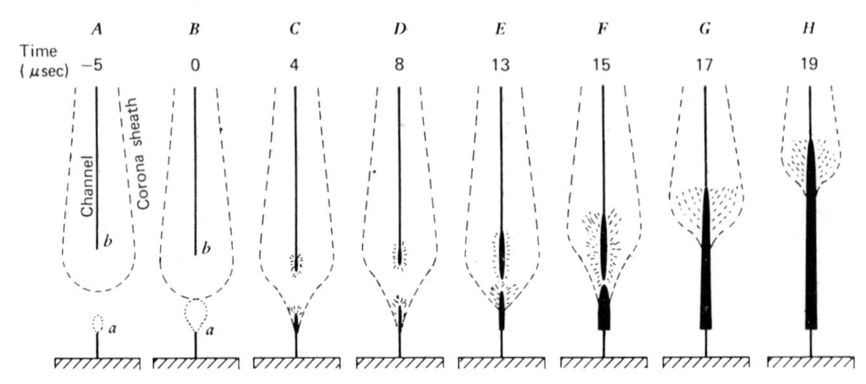

Fig. 9.3. *Stages in the development of an upward channel. After Wagner.*[3]

the leader head and by the return streamer from the stricken tower, as indicated by stages *D*, *E*, and *F*, is a factor in decreasing the velocity of the return streamer and in raising the effective impedance of the leader path. According to Wagner,[3] this impedance is on the order of 3000 Ω over the last 100 m between the point of discrimination and the stricken object. The effect on the current wavefront is to produce a concave upward shape, representing the growth of the current from the charge-collection process. When the two main channels, meet the discharge current is at maximum value (i.e., its curve flattens out).

The subsequent dart leaders have a higher velocity than the original, and consequently the stroke-current wavefronts are somewhat steeper.

From an analysis made by Wagner,[3] the parameters for two typical strokes are listed in Table 9.1. This indicates that the surge impedance

TABLE 9.1 PARAMETERS OF TWO LIGHTNING STROKES[a]

Current	i (A)	50,000	10,000
	D (ft)	300	300
	b (ft)	6	2
	v, a numeric	0.3	0.17
Potential	V (kV)	46,000	20,000
Impedance	Z (Ω)	920	2000
	Striking distance (ft)	310	135

[a] From Wagner.[3]

D = length of (return) stroke when current has reached nominal crest value, i.

b = assumed radius of return stroke channel.

v = velocity of propagation relative to that of light.

V = stroke potential at current, i.

Z = surge impedance of return stroke at current, i.

increases with shortening of the striking distance (i.e., as the head of the return stroke rises), which contributes to the fall in stroke current following the crest value.

9.3 THE OVERALL DISCHARGE PATHS

The stepped leader preceding a return stroke might strike the transmission tower, the overhead "ground" wire, the conductor at midspan, or the earth at some distance from the line, depending on the position of the thundercloud and the intensity of its charge. Even a phase conductor, on a line with overhead ground wire, can be struck.

The different paths by which a transmission line can be struck by lightning have been well illustrated by Wood,[4] whose sketch is reproduced in Fig. 9.4

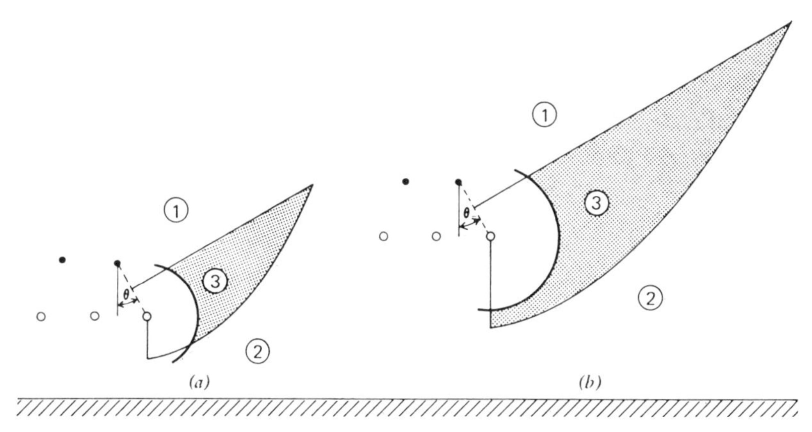

Fig. 9.4. *Earth-wire shielding—geometrical model for (a) low and (b) high towers. Key: ●, earth wire; ○, phase conductor; θ, protective angle. After Wood.*[4]

for both low and high towers. There is a direct path from the capacitive charged head of the leader stroke to the overhead "ground" conductor of the transmission line (1). The tower length presents a surge impedance to the current, which, combined with the tower earth resistance, develops a high potential at the tower top. This voltage can flash across insulator strings to the phase conductors, that is, a backflashover may occur. Second, there is an indirect path (2) where a primary discharge occurs between the capacitance of the charged leader head and earth, at some distance from the line. The electrical field produced by the high-current return stroke may induce a high voltage surge on the line conductors, sufficient to arc across the insulators. This kind of strike is usually called an induced voltage due to a nearby stroke to earth.[5] Line faults due to indirect strokes are generally limited to lower voltage transmission lines, where lower towers and shorter insulator strings are used. For higher towers a strike to the earth near the tower base could raise the potential of the tower to such an extent that a backflashover would occur from tower to live conductors.

The third path is between the leader head and a phase conductor, forcing a high current into the conductor whose surge impedance raises the potential of the conductor, sufficiently in some cases to flash over to the tower. This is the "shielding failure" type of discharge because the overhead ground conductor fails to shield the line.

The geometry of Fig. 9.4 helps to illustrate the relative vulnerability of low and high towers, via the three different paths. The straight-line perpendicular to the line joining "ground" wire and phase conductor at its midpoint is equidistant from these two conductors. Stepped leaders approaching from above this line in region 1 will strike the overhead "ground" wire. Between the "equidistant" line and the earth plane, a long curved line is drawn that is equidistant from the phase conductor and the earth plane. The line and curve enclose area (3) where the phase conductor is vulnerable to lightning strikes, which may flash across the supporting insulator to the tower. An arc of a circle centered on the phase conductor can be drawn to represent an area within which the lightning voltage would choose the tower as a terminal rather than the insulated phase conductor. In area 2 between the ground plane and curved line, the earth would receive the strike and at times would raise the tower potential to a value that would cause a backflashover between tower and phase conductor. The figure illustrates that area 3 is much larger for the higher tower, making the phase conductor more vulnerable to strikes.

To reduce the size of area 3 the protective angle θ must be made smaller by moving the overhead "ground" wire outward. That is to say, taller towers require a smaller protective angle for a given degree of protection. According to Wood,[4] the protective angle should be 45, 25, and 10 degrees for tower heights of 50, 100, and 150 ft, respectively, on ehv transmission lines. Another authority, Wagner,[3] has arrived at similar values—that is, 45, 30, and 12 degrees—based on data from analysis and operating experience.

9.4 GENERAL DESIGN FACTORS

The effect of lightning on a transmission line is influenced by some general design factors. If, for example, a line carries two circuits, there is an advan-

Fig. 9.5. *Pathfinder device; it can differentiate between a backflashover and a shielding failure. From Wood.*[4]

tage in having the insulation value of one different from the other. With this arrangement a backflashover will cause a fault on only one of the lines rather than on both.

On dc transmission lines a conductor of one polarity only, usually positive, will be struck at a time.

A trend in transmission-line design is fast-acting autoreclosing switchgear, which can minimize lightning outage time.

One of the newer devices for identifying lightning faults on transmission lines is shown in Fig. 9.5. This pathfinder device can indicate three different events: a backflashover from tower to phase conductor, a flashover in the opposite direction from phase conductor to tower, and a power arc if it follows the flashover. Its value is in determining the proportion of strokes that are not diverted by the lightning protection system of overhead "ground" wire and buried electrode system.

9.5 THE AIR TERMINAL

The established method of providing a lightning-discharge terminal on power-transmission lines is the overhead "ground" conductor. One or more of these conductors are installed so as to provide protective angles of approximately 45, 30, and 12 degrees for tower heights of 50, 100, and 150 ft, respectively. Considerable amounts of data have been compiled on shielding failures versus protective angle,[3] and a representative curve is shown in Fig. 9.6. Shielding has increased importance for high-voltage lines. According to Golde,[1] flashovers on 275-kV lines are mainly due to shielding failures, whereas on 132-kV lines the failures are due principally to backflashovers.

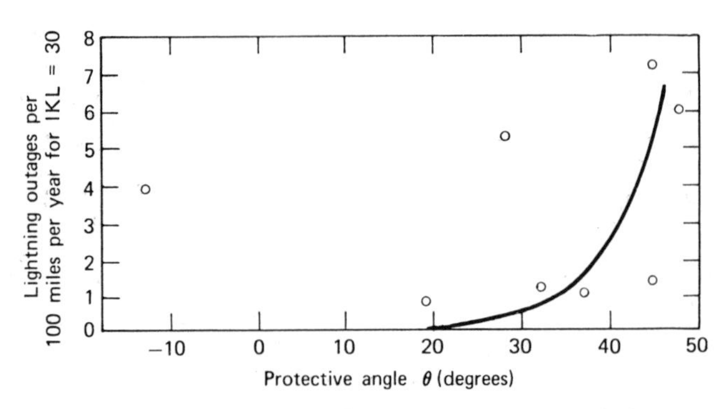

Fig. 9.6. *Lightning-induced outages as a function of protective angle for 110 to 165 kV, with average tower-footing resistances of less than 10 Ω. IKL: isoceraunic level. From Wagner.*[3]

9.6 TOWER IMPEDANCE

It has been determined with some degree of certainty[3] that the tower surge impedance looking into the tower top comprises a relatively small portion of the total impedance facing the lightning discharge. The surge impedance is difficult to calculate because of the tower's nonuniform shape, its lattice-type construction, and its relationship to earth. Analyses and measurements by several investigators have yielded different results.

By conventional methods of calculation the surge impedance of a 100-ft tower with an inductance of 20 μH would be about 200 Ω to a current wavefront of 4 × 40. However, this current wavefront entering the top of the tower would produce a voltage equal to the product of the inductance and the rate of change of current, that is, $L\,di/dt$. If the wavefront rises to a crest value I in 4 μsec and the tower inductance is 20 μH the voltage developed would be $I \times (10^6/4) \times 20 \times 10^{-6}$, or $5I$ volts. That is, the apparent impedance of the tower is 5 Ω. As a usual value for tower-footing resistance is 20 or 25 Ω, the 5-Ω tower surge impedance is only 20% or less of the total impedance to earth. Stating the matter in another way, the effectiveness of a ground system is not much derated by the surge impedance of the tower.

9.7 THE BURIED EARTH SYSTEM

A conventional buried electrode for power lines is a buried earth wire called a counterpoise running below the line, along its length. Looking from the top of any tower the impedance of this buried conductor running in opposite directions from the tower footing would be that of two buried wires in parallel. In addition there would be a small capacitance between the overhead ground conductor and the counterpoise. The footings of the towers would constitute some resistance in parallel with the counterpoise. In locations of very low ground conductivity but with a layer of top soil additional buried radial conductors can be installed.

It would be advisable and relatively inexpensive to provide "cages" in the tower-footing excavations, as described in Section 7.3. Let us assume four footing excavations per tower, each 6 ft deep and 6 ft square. Putting a vertical conductor in each corner and bonding these with a horizontal ring conductor would yield a resistance to earth of about 75 Ω in soil material with a resistivity of 2000 Ω-m. Four such cages in parallel separated by more than 6 ft would have a combined resistance of 25 Ω. This system in parallel with the counterpoise would have a resultant impedance of about 15 Ω. Adding 5Ω for the apparent surge impedance of the tower gives a resultant value of 20 Ω.

If one has a tower-earth-system impedance of, say, 20 Ω and the insulator strings will withstand 10^6 V, protection would be afforded for all lightning discharges to the tower tops or overhead "ground" wire of 50 kA or less. For the isoceraunic conditions of Fig. 9.2, where five strokes occur per 100 miles of line per year and 5% of these exceed 50 kA, one could expect one lightning fault per 100 miles of line every 4 years.

This example illustrates that a lower tower ground resistance requires less insulation of the live conductors for a given degree of lightning protection.

This relationship is shown in the probability curves[6] of Fig. 9.7, which indicate the probability of flashover as a function of stroke current and the number of insulator units in a string. Each probability curve has the corresponding value of tower-ground resistance marked on it. These curves, which are drawn for an isoceraunic level of 30, are valuable in determining the economical design parameters for any isoceraunic level.

9.8 PROTECTIVE LEAKAGE PATHS—PIPE–PIPE GAPS

Attention is now being directed toward bleeding off pre-discharge leakage currents produced by the high voltage gradients ahead of a stepped leader. This current leakage occurs between the overhead ground wire and the phase conductors of the transmission line. The existence of this leakage path lowers the potential gradient caused by an approaching leader tip, or stated another way, the line withstands a higher potential produced by a stepped leader before flashing.

A device to increase the leakage current and thereby increase the protection to the power line from lightning is called a pipe–pipe gap because pipe sections were first used to form the gap. As indicated in Fig. 9.8, the lengths of pipe are mounted parallel to each other, and the gap between them is made just greater than the length of the insulator string. The lengths of the pipe sections are limited only by structural requirements. Such gaps, by permitting high current leakage under thunderstorm conditions, can hold off lightning strikes until the potential gradient rises significantly above the value that would otherwise permit a lightning stroke to occur.

Another device to protect transmission lines is the tube protector,[6] which consists of a gap within an insulating tube, normally mounted below a live conductor. The rating of the tube is chosen so that its gap will flash over before the insulator breakdown voltage is reached. The impedance characteristics of the tube enable it to regain its insulating property quickly while the discharge voltage is going through a minimum of its oscillation. The protector tube is applicable to ac lines.

9.9 PROTECTION OF SUBSTATIONS

Although power-supply agencies usually provide lightning protection for their equipment, incidents of equipment damage due to lightning still occur. An example of such damage to a high-voltage transformer is shown in Fig. 9.9.

For the protection of substations or switching yards the three fundamentals of air terminal, downlead, and earth system are, of course, required.

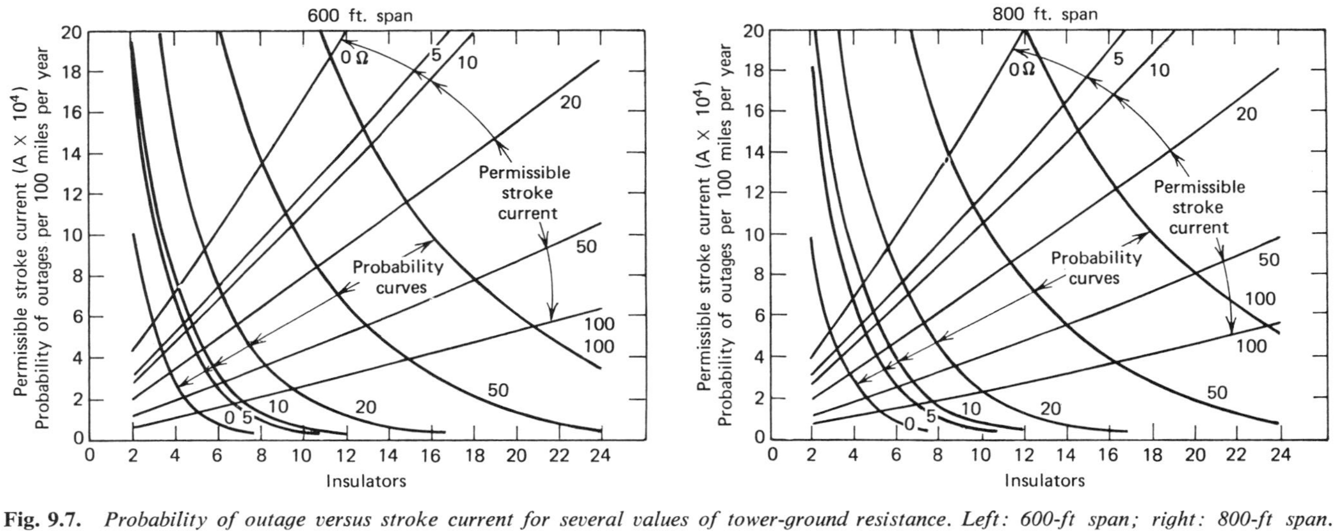

Fig. 9.7. *Probability of outage versus stroke current for several values of tower-ground resistance. Left: 600-ft span; right: 800-ft span. From Ref. 6.*

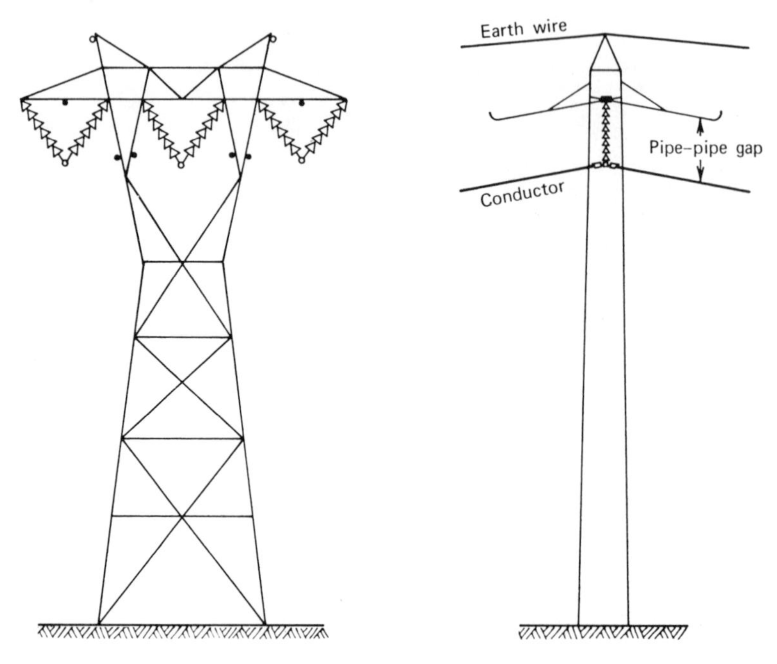

Fig. 9.8. *Application of pipe–pipe gaps to transmission lines. After Wood.*[4]

The air terminal is usually several overhead wires strung between the switching frames or towers, vertical rods mounted on top of the steel towers or frames, or separate masts on the perimeter with wire strung between their tops.

The down conductor is normally provided by the legs of the metal switch frames or towers, and from the base of these structures a number of heavy leads connect to a buried ground system of horizontal conductors and rods. An effective coupling with the earth can be achieved by the grid system described in Section 4.15. As there is usually some discretion permitted in locating the station, an effort should be made to find a site with deep surface soil and to avoid bare-rock surfaces.

In at least one large station in the tropics the lightning protection system is kept separate from the power ground system.[4] This reduces the possibility of raising the power-ground-system potential sufficiently to cause back-flashovers to the equipment. The lightning protection is provided by several tall masts placed on the perimeter of the station, each having buried radial wires emanating from its base. The buried conductors are independent of

Fig. 9.9. *High voltage transformer damaged by lightning.*[7]

the ground conductors for the power apparatus. The lightning masts are made tall enough for their protective zone to encompass the equipment area.

For conventional substation grounds of buried earth electrodes the earth can be temporarily raised to a high potential by a lightning discharge. Therefore any auxiliary service cables buried underground require high insulation between sheath and core to prevent a voltage breakdown. However, if there are buried radial wires lying above the cable but connected to ground rods (i.e., having a low impedance to earth), they will protect the cable to a high degree.

A grounding system for a small substation is illustrated in Fig. 9.10.

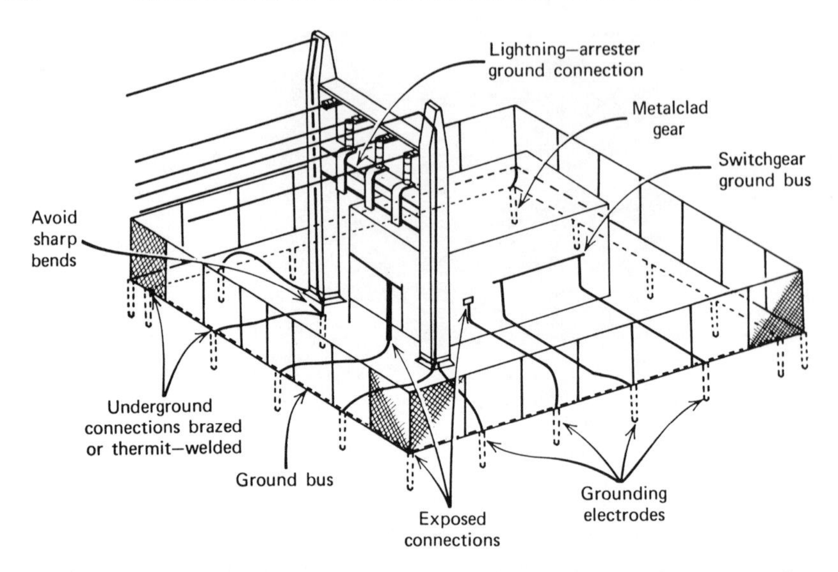

Fig. 9.10. *Typical grounding system for an outdoor substation. After Carpenter.*[8]

9.10 UNDERGROUND CABLES

The impedance of the sheath of an underground cable to a current surge is on the order of 100 Ω and falls off to the ohmic earth resistance of the conductor as the surge dies out.[9] It has also been found that attenuation of the current wave pulse along a buried conductor is rapid, so that it diminishes to a low value in a length of 100 m. Voltage reflections are therefore not a serious problem. But the magnitude of the induced voltage on a cable sheath by a current pulse is on the order of 50 kV/kA of entering current.[9] It follows that in lightning-prone areas cables buried in soil of low conductivity require high insulation or protection by overlying buried earth conductors, associated with ground rods if necessary, to provide an earth resistance that is relatively low compared with the cable sheath itself.

9.11 LIGHTNING ARRESTERS

The lightning arrester is a "safety valve" used to limit overvoltages and to bypass the associated current surge to a ground system. The arrester is the triggering device, but the ground system absorbs most of the energy. The purpose of the arrester, or surge diverter, is to protect electrical equipment from being damaged by voltage surges and to minimize the interruption of electrical systems. Consequently they are normally installed adjacent to the transformer, reactor, or other component needing protection; the sparkover voltage rating of the arrester is selected in relation to the normal voltages of the system to which they are connected.

The arrester has a dual role of protecting against abnormal overvoltages due to faults on the system and protecting against voltage surges due to lightning. Accordingly its threshold or sparkover voltage should be nearly constant for overvoltages at power frequency, switching transients at medium frequencies, and lightning surges that are steep-fronted or high-frequency impulses. Also the arrester should discharge at a relatively small margin of voltage above the normal maximum voltage of the protected system, so that the insulation requirements of the transmission line and connected components can be kept to a minimum. It follows that the arrester should restore itself for a repeat operation when the overvoltage drops to near normal level; that is, it should have a low "reseal" voltage.

There are three general classes of lightning arrester corresponding to three voltage ranges: (a) distribution type, with voltage ratings of 1 to 20 kV; (b) secondary, or line, type, with ratings in the 20- to 75-kV range; and (c) station type, with ratings from a few kilovolts to the highest transmission-line voltages used. The station-type arresters have more advanced design features, better discharge characteristics, and greater reliability.

The present-day arresters comprise current-limiting gaps in series with valve elements, such as silicon carbide blocks. To proportion the surge voltage equally across the series gaps, they are shunted by nonlinear grading resistors. Grading capacitors are also used across the gaps in the more advanced designs to level out the voltage versus rise time or frequency characteristic. These capacitors supplement the inherent shunt capacity of the arrester. A cross-sectional view of high-voltage-arrester construction is shown in Fig. 9.11, and the relationship between the sparkover voltage and the rise time of the voltage surge is shown in Fig. 9.12. Curves B and C in Fig. 9.12 represent new arrester types with grading capacitors, while curve A is for an older type without a good grading structure.

In operation, a current-limiting gap strikes at a predetermined voltage, and then its arc is forced into a longer path by a magnetic field effect. This raises its arc-drop voltage, which tends to extinguish the arc and prevent the flow of power follow current. The valve elements, having a negative resistance characteristic, present a low impedance to the initial high voltage surge, but when the voltage drop across them is lowered due to a higher arc-drop

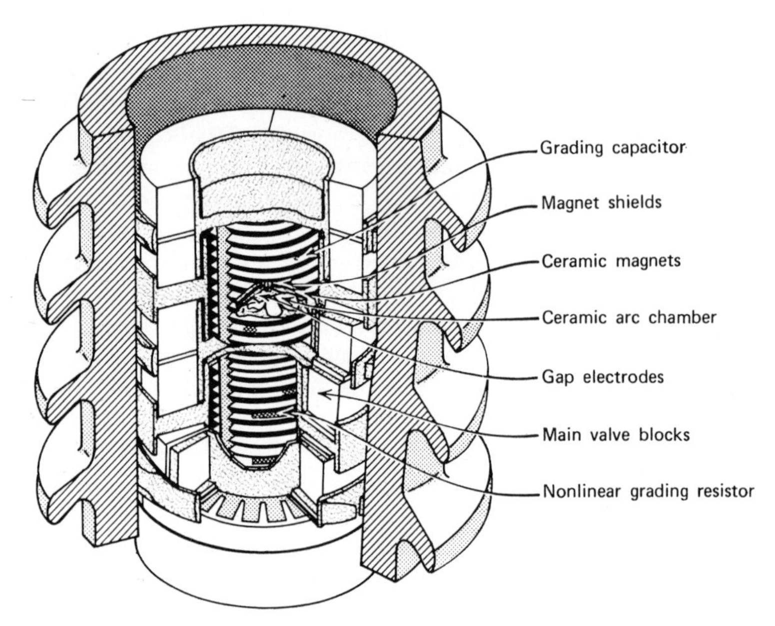

Fig. 9.11. *Cross section of current-limiting arrester. After Vogler.*[10]

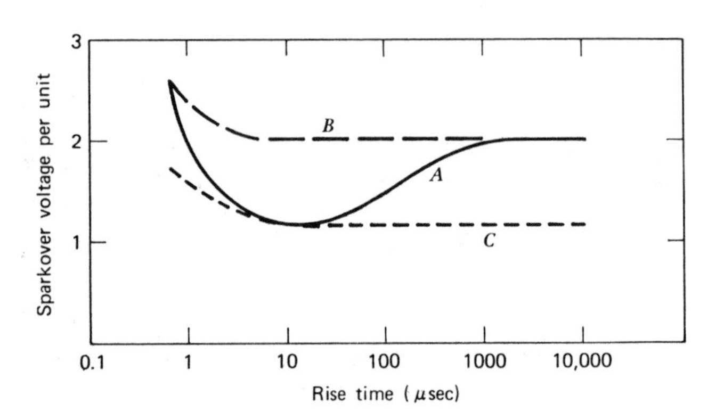

Fig. 9.12. *Arrester sparkover response as a function of surge-voltage rise time. After Yost.*[3]

voltage across the series gaps, the impedance of the valve blocks increases, further impeding the flow of power follow current.

The discharge characteristics of three current-limiting gap-type arresters are illustrated in Fig. 9.13. The most efficient gap is in type *a*, where the gap-voltage peak rose to the highest value and the power follow current was limited to the lowest value.

The rapid arc-extinction action of the current-limiting gap-type arrester permits it to restore itself to operating condition, or reseal, at a relatively small margin above its sparkover value.

The ability of insulation to withstand overvoltage depends on the duration of the surge: the shorter the impulse, the higher the so-called withstand voltage. Accordingly insulation has one value of impulse withstand voltage for switching surges of relatively long duration and a higher value for lightning impulses. The value of insulation on electrical apparatus is coordinated with the maximum operating voltages expected, and the arresters are selected

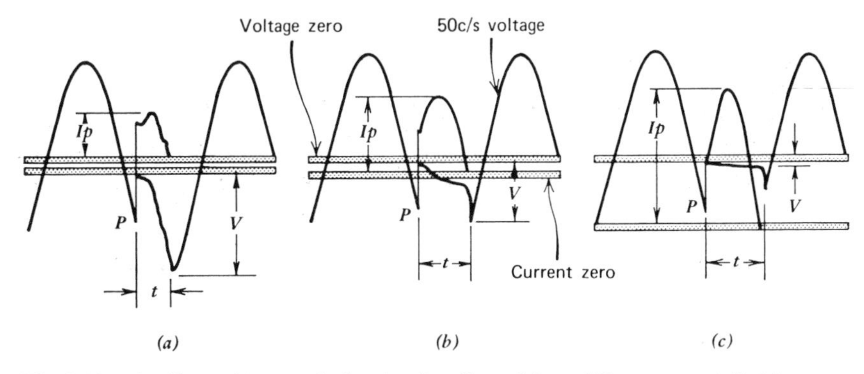

Fig. 9.13. *Oscillographic records showing the effect of three different current limiting gaps. Oscillogram* (*a*) *was taken with very efficient magnetic arc control device. The gap used for oscillogram* (*b*) *was less efficient and* (*c*) *was the least efficient. The parameters are given below and it will be seen that the lowest follow current occurs where the gap voltage peak is highest. All tests were carried out using the same circuit at 3 kV rms.*

Symbol	Parameter	Units	Value (a)	(b)	(c)
Ip	Follow current peak	A	66	100	172
P	Point of arc initiation	degrees	25	25	25
t	Time of arc duration	ms	4.2	6.7	7.2
V	Gap voltage peak	V	2,880	680	170

to discharge at voltages of about 1.5 times the normal maximums expected. The present-day arresters will reseal at about 1.3 times their voltage rating. The voltage rating of an arrester is numerically less than the withstand voltages it protects. For example, in a high-voltage system an arrester rated at 612 kV might be used on a 735-kV system where it would be required to flash over on a switching surge impulse of 1230-kV crest and to flash over on a lightning impulse of 1400-kV crest.

The relationship between equipment voltage ratings and the corresponding withstand voltages for high-voltage systems is shown in Table 9.2,[12] where the rated switching impulse voltage is taken as $\sqrt{2}/\sqrt{3}$ times the highest operating voltage of the equipment. The switching impulse withstand voltage is obtained by multiplying by the overvoltage factor, which is listed under the equipment voltages. The lightning impulse withstand voltage is nominally taken as three standard deviations[12] above the maximum switching impulse voltage. The arresters would be selected to flash over at the switching surge

TABLE 9.2 INTERNATIONAL STANDARD INSULATION LEVELS[a]

Highest Voltage for Equipment (U_m kV, rms)					Switching Impulse Withstand Voltage (kV Peak)	Lightning Impulse Withstand Voltage (kV Peak)
300	362	420	525	765		Apparatus Protected by Surge Diverters
3.45	2.86				850	950–1050
	3.20	2.76			950	1050–1175
		3.06	2.45		1050	1175–1300–1425
			2.74		1175	1300–1425–1550
Rated switching impulse voltage: $\frac{\sqrt{2}\, U_m}{\sqrt{3}}$				2.08	1300	1425–1550–1800
				2.28	1425	1550–1800–2100
				2.48	1550	1800–1950–2400

NOTE: The standard series of voltages applicable to both switching and lightning impulse voltages is as follows (kV):

850, 950, 1.050, 1.175, 1.300, 1.425, 1.550, 1.675, 1.800, 1.950, 2.100, 2.250, 2.400, 2.550, 2.700, 2.900.

When, due to the design of the system or the methods for control of lightning or switching surges, combinations other than those shown in this table are economically and technically justifiable, the values shall be selected from the series.

[a] From Guertin and McGillis.[12]

voltage (medium-frequency) and lightning surge voltage (steep-fronted, high-frequency) as given in the center and right-hand column.

The energy stored in a high-voltage transmission line is large compared to the thermal capacity of an arrester. This energy varies in magnitude as the square of the line voltage, is proportional to the length of the line section affected, and is inversely proportional to the surge impedance of the line. On the other hand, the thermal capacity of an arrester is only proportional to its voltage rating (i.e., to its size). This emphasizes the importance of the grounding system to absorb the discharge energy. The current-limiting gap-type arrester absorbs a relatively small proportion of the energy passing through it; this makes it effective in withstanding repetitive action, such as successive flashing across dirty insulators or multiple lightning strokes.

Arresters of the current-limiting gap type perform satisfactorily on dc transmission lines, whereas arresters of earlier design, employing valve elements only, depended on the current wave going through a "zero" to extinguish the arc—that is, they were suitable for ac lines only.

An illustration of three high-voltage arresters installed to protect equipment at a power substation is shown in Fig. 9.14. It will be noted that they have capacity or grading rings mounted near their tops.

The application of arresters, including their optimum point of connection relative to the apparatus they protect, is a broad subject; so only an outline has been given here to indicate its relationship to lightning protection.

REFERENCES

1. R. H. Golde, "Lightning Performance of High Voltage Distribution Systems," *Proc. IEE,* **113,** No. 4, April 1966.
2. "A Method of Estimating Lightning Performance of Transmission Lines," Report of the AIEE Transmission and Distribution Committee, *Trans. AIEE,* **69,** 1950.
3. C. F. Wagner, "Lightning and Transmission Lines," *J. Franklin Inst.,* **283,** No. 6, June 1967.
4. A. B. Wood, "Lightning and the Transmission Engineer," *Electronics and Power,* **15,** 1969.
5. P. Chowdhuri and E. T. B. Gross, "Voltage Surges Induced on Overhead Lines by Lightning Strokes," *Proc. IEE,* **114,** No. 12, December 1967.
6. Engineers of Westinghouse Electric Corporation, *Electrical Transmission and Distribution Reference Book,* 4th ed., 1964, Chapters 16 and 17.
7. "Lightning Performance," *IEE News (London),* December 9, 1968.
8. L. J. Carpenter, "Equipment Grounding for Industrial Plants," *Electrical Engineering,* March 1954.

Fig. 9.14. *Three high-voltage arresters with grading rings. Courtesy of Canadian Ohio Brass.*

9. Einhorn and Goodlet, "Lightning Overvoltages on Underground Cables," *Proc. IEE,* **88,** Part 2, 1941.

10. Vogler, *Electrical News and Engineering* (*Toronto*), June 1968.

11. A. G. Yost, "Lightning Arresters for EHV," *Electrical News and Engineering* (*Toronto*), December 1966.

12. M. B. Guertin and D. McGillis, "Effects of Overvoltage Distribution on External Insulation in EHV Systems," *Electrical News and Engineering* (*Toronto*), July 1971.

Index